Alfred Crofts

How to make a Dynamo

A practical Treatise for Amateurs

Alfred Crofts

How to make a Dynamo
A practical Treatise for Amateurs

ISBN/EAN: 9783337106621

Printed in Europe, USA, Canada, Australia, Japan

Cover: Foto ©ninafisch / pixelio.de

More available books at **www.hansebooks.com**

HOW TO MAKE A DYNAMO:

A Practical Treatise for Amateurs.

Containing numerous Illustrations and detailed Instructions for Constructing a small Dynamo,

TO PRODUCE THE ELECTRIC LIGHT.

BY

ALFRED CROFTS.

THIRD EDITION. REVISED AND ENLARGED.

London:
CROSBY LOCKWOOD & SON,
7, Stationers' Hall Court, Ludgate Hill, E.C.

1890.

[All Rights Reserved.]

PREFACE TO THE THIRD EDITION.

The first and second editions of this little book being so rapidly disposed of; the sale of over 2,000 copies is sufficient evidence that a want existed for such an inexpensive little work, not only by amateur electricians to whom the treatise was originally addressed, but by professed engineers and mechanics at home and abroad, who have by its aid made their own dynamos to light their dwellings or workshops.

The patrons of the former editions included amateurs of rank as well as hard working mechanics, while several letters expressive of delight have reached the Author, for which he takes the present opportunity of returning his thanks, and for the support accorded to the two past editions.

That the instructions were carefully worked out by experiment before being printed, an extract from an Engineer's letter appearing on page 11 will confirm.

During the run of the first and second editions many tons of castings for working these little dynamos have been turned out at the Phœnix Foundry in Dover under the Writer's supervision, and have been sent to America, Australia, New Zealand, Switzerland, &c.

It is most desirable, however, that the iron employed should be *soft*, otherwise the efficiency of the generator would be affected; therefore, to obtain satisfactory results, it is imperative to secure first class castings.

In the Appendix will be found some useful additional particulars of a Dynamo with wrought iron field cores, the shoulders of which can be easily turned up in a small lathe, and by employing a gramme ring armature and commutator as described will easily light seven twenty-candle-power lamps to their full brilliancy.

With these remarks "HOW TO MAKE A DYNAMO" enters upon its THIRD EDITION.

7, CLARENDON PLACE, DOVER,
March, 1890.

HOW TO MAKE A DYNAMO.

CHAPTER I.

WHILST the increasing numbers of professed electricians have during recent years greatly aided in developing the use of electricity, there is yet very much good work done by a considerable body of students in this science outside the rank and file, who, with no expectation of fee or reward for their labour—beyond the gratification it affords them—work hard, both mentally and physically, in scheming and manufacturing some electrical instrument or machine—they form, both at home and abroad, a large volunteer force of amateur electricians. Since the more general application of electricity to the purposes of illumination, this branch of electrical engineering has recruited its ranks with many who possess a scientific turn of mind, and welcome every opportunity of gaining some insight into the interesting and fascinating subject of Electric Lighting.

The amateur's productions may sometimes be despised by their more accomplished professional brethren; yet frequently scientific models may be seen at the Industrial Exhibitions and Schools of Engineering, of a nature sufficiently encouraging indeed to stimulate other novices to imitate, and considering the crude tools and appliances by which such productions are often accomplished, they frequently show workmanship that is highly creditable to the makers.

But there are two classes of amateur electrical engineers, just as among experimenters in other branches of science, viz. :—the one with a well-filled purse, and the other with but a few shillings at his command; and yet how often the latter with his scanty assortment of tools, will turn out the best work! There is yet another disadvantage that some amateurs have to deal with, viz. :—the want of time for working; perhaps only a few minutes can be snatched in the dinner hour, or after the day's toil in some other occupation is over.

Many of the pioneers of electricity were but amateurs in the true sense of the word; men who had strayed from the path of their legitimate calling. A telegraph instrument which still bears the inventor's name—the Morse printer—was the outcome of an amateur's brain. Samuel Morse was an artist by profession, but was attracted into the path of science, and left the painter's easel to pursue the study of electricity, and it was on board ship, with all the disadvantages of scanty and inferior workshop appliances, that this dabbler in electrical science conceived his brilliant idea, and worked out the models of his recording telegraph, an instrument which holds a leading position in telegraphy to the present day in binding the world together in friendly intercourse.

The illustrious Faraday, whose portrait appears in the *Frontispiece*, did not start life in a business associated with electricity in any way; as he was apprenticed to a bookbinder, but an attendance upon some lectures by Sir Humphry Davy drew his enquiring mind to the study of electricity. With a few pieces of sealing wax, some copper wire, and sundry pieces of iron and steel, he made those startling discoveries by which he has left as a monument of Fame to his memory, the knowledge of those laws of nature whose secrets he laid bare in all their surprising beauty. To one not articled or apprenticed to science, but who followed her sincerely as a devoted servant, is due the unfolding of those hidden laws in magnetism and electricity which now guide the electrician of the present day in the work which he performs.

The amateur electrician possessing even the most scanty stock of tools, may nevertheless be able to construct many useful electrical articles, (such as a galvanometer, induction coil or battery) without the complete workshops of his more wealthy brethren; for it is surprising what an insight may be gained by the manufacture of such appliances, and their usefulness may be measured by the fact that Franklin demonstrated the identity of lightning with electricity by means of a sheet of brown paper, a ball of twine or silk thread and an iron key. Another illustration of an amateur is to be found in Arkwright, who perfected his invention of the spinning machine in the uncongenial atmosphere of a barber's shop; and with the further disadvantage of a wife who had a persistent dislike to his designing machinery, and once smashed his models on the very eve of their completion. She frequently rebuked him for neglecting his lucrative occupation of shaving customers, and often supplemented her censure with "cuss the cheenery."

A further example of the value of an amateur may be observed in the father of railways, George Stephenson, inventor of the locomotive, who developed his remarkable engineering skill in the obscurity of a coal mine, adding to his slender earnings by mending a neighbour's clock or watch.

It is also no secret that Edison, the clever inventor of modern times, commenced his business career as a newspaper boy on a railway in America, and some of his early experiments were conducted in a corner of the break-van during such moments that could be spared during intervals of the sales of newspapers.

The foregoing instances of celebrated amateur's skill and inventive genius are noted to show that great results may spring from early tastes displayed by the tyro in science, and in order that the reader, especially if he be juvenile and with a fancy for electrical engineering, may not be discouraged by inconveniences, both mechanical and financial, as they arise. With some bold minds it appears quite immaterial how great the apparently insuperable obstacles may be with which they are confronted, their determination to succeed generally leads them to triumph over all difficulties.

Many amateurs of the present day are the happy possessors of a lathe, (the most enviable tool for all young aspirants to electrical engineering,) and the owner of say a 4½ in. or 5 in. centre back geared lathe with slide rest, will find such an appendage to his workshop extremely useful in the construction of the machine these instructions refer to, viz.: a serviceable little dynamo which so many tyros are ambitious to rig up for themselves, and which, when completed, will afford complete satisfaction to the maker, as it will enable him, by means of his own skill, to display an installation of the electric light, as the following extract from a letter received by the Author, will confirm:—

<p style="text-align:center;">METROPOLITAN ASYLUM DISTRICT,
Darenth, near Dartford, Kent, <i>Aug. 5th</i>, 1889.</p>

Dear Sir,

I have very great pleasure in informing you, that by means of your admirable castings, and explicit book on "*How to make a Dynamo*," I have satisfactorily constructed a 120 candle power dynamo. I had it running for four hours after I finished it, with six of Edison's swan 46 volt lamps, to their full brilliancy; and I must tell you that an electrical expert, Mr. Dale, whom I know, said it was all that could be desired.

<p style="text-align:center;">Yours &c.,
HENRY GILES, <i>Engineer</i>.</p>

To Mr. Alfred Crofts,
    Electrician, Dover.

CHAPTER II.

As a general rule the amateur pays rather dearly for the materials required in carrying out his hobby; and it is seldom that the tyro in mechanical matters can dispose of his productions with any advantage to himself, but, on the contrary, with a certain loss of money, patience and time; therefore, in a financial point of view, it would be false economy for him to attempt the manufacture of a set of models for a single dynamo, as the patterns require to be made with the greatest exactness and finish in order that the resulting castings from them should be perfect in every respect. Pattern making, like moulding, is a skilled branch of its own, and as the Dover foundry is now making a speciality of soft iron castings of small dynamos, which are obtainable by the amateur ready to hand, he can be fairly started with the necessary substantial materials at a moderate cost, and, apart from the knowledge that will be gained by building his own dynamo, it will, if carefully made, leave a margin of remuneration for the labour bestowed upon it, as the requisite materials can be bought for at least one-fourth of what it would cost to buy the finished article. By following a machine of good repute there is no costly experience to purchase in making trials, as amateurs often do, upon vague ideas of their own, nor the risk of disappointing results, yet one word of caution is necessary—do not be in too much hurry, as sometimes an amateur's zeal is apt to confirm an established maxim " the more haste, the less speed."

Before describing the type of dynamo to be selected, it will be worth while to take a retrospective glance at the early history of the electric light, and its progress through the employment of magneto electric generators and the modern dynamo.

The electric arc was first displayed at the Royal Institution by Sir Humphry Davy in the year 1801. He employed a large number of galvanic cells, from which two wires were led to a couple of charcoal pencils: these pieces of charcoal were brought together and afterwards slightly separated, when a brilliant light was emitted from the ends thus almost touching.

At this stage, however, the electric light could only be regarded as a wonderful scientific toy for the philosopher, since the troublesome and costly method of generating electricity by the chemical action of batteries prevented its use from becoming general for illumination purposes, and for a long time it was confined to the lecture hall and stage, until, in the year 1831, the time arrived when nature whispered her secret of the laws of

MAGNETIC INDUCTION

into the ears of Michael Faraday, the son of a

village blacksmith. Young Faraday had become the pupil of Sir Humphry Davy, who perceived qualities in the country blacksmith's son which indicated that he was better adapted for scientific research than in following the trade of a bookbinder, which he had commenced learning. Professor Davy afforded the lad every opportunity of gratifying his particular taste, and the young philosopher pursued the study of science with such devoted love that nature bestowed upon him her reward; for she selected him as the man of genius to unfold her hidden mysteries, and to proclaim to an astonished world how currents of electricity could be generated in a coil of wire when suddenly brought near, and quickly removed from the poles of a magnetised bar of iron or steel. This brilliant discovery was the starting point which ultimately led to mechanical energy being converted into light by means of

MAGNETO ELECTRIC GENERATORS,

with which in the early Alliance machines constructed by Nollett, of Brussels, and subsequently improved by Holmes, of England, some of the first important experiments in electric lighting were carried out at the South Foreland lighthouses, St. Margarets, near Dover. The magneto generators are still in use at these lighthouses, and are constructed upon the principle of Faraday's discovery of magnetic induction by which coils of insulated copper wire are rapidly rotated near the poles of fixed steel horse-shoe-shaped permanent magnets; the rapid revolution of these coils of wire produces alternate currents of electricity; the exciting magnets are arranged in fixed circular form with their **N** and **S** poles placed alternately, and separated by equal spaces; the moving coils of insulated wire are placed at regular distances upon a revolving drum, and constitute what is termed an armature; when these coils in their rapid revolution approach the pole ends of the stationary steel magnets, a momentary current of electricity is induced in one direction, and upon the coils receding from these exciting field magnets, an instantaneous current is also established, but in an opposite direction to the former one. Machines of this type are known as *alternating* current generators, and it may be remarked that in electric arc lamps worked on this system the upper and lower carbons burn away equally,—an important consideration in lighthouses, since the focus of the arc can be maintained at one fixed point.

A series of alternating currents are developed in a single revolution of the armature, the changes of current being dependent upon the number of magnets passed by the coils in their circular path of rotation. De Meritens, of France, has also made improvements in magneto machines, both as regards reduced size and slower speed; the coils of wire in the De Meritens' armature are of ring form, but built up in annular shaped sections. In this, as in Holmes', there is no commutator required, since the two extremities of the entire wire covering the ring furnish the alternating current for the arc. Magneto electric generators, however, excepting for lighthouses and other special purposes, have been superseded by

DYNAMO ELECTRIC MACHINES,

for based upon the laws of magnetic induction, modern science has developed generators known as dynamo machines, in which the steel exciting magnets are dispensed with, and the iron carcase of the machine is wound with insulated copper wire in such a manner as to form two electro-magnetic poles, one of north and the other of south polarity, when a current of electricity is caused to flow through the wire; iron does not retain magnetism like steel, although it is susceptible of a higher degree of magnetic power, and the horn-shaped pole pieces of a dynamo which embrace the armature, being of cast iron, will contain some residual magnetism after the current from a galvanic battery has been passed through the wires surrounding the limbs of the field magnets which are connected to the pole pieces. The magnetism remaining in the poles is but weak, yet is sufficient to induce a current of electricity to flow in the wire of the armature, when it is quickly revolved between them. The current thus generated in the armature is conducted through the wire system of the field magnets, thus augmenting their exciting power, and consequently increasing the current first induced in the armature, which again re-acts upon the field magnets in a much stronger degree, and they in turn excite an intensified current in the armature, and so on, until a powerful current is obtained. This is the principle of a dynamo of which the Gramme (Fig. 1) is a

Fig. 1.

leading type, for after the patent had expired in 1884, makers sprang up all round to manufacture it, either in the original form, or with certain questionable improvements, which confirmed the high reputation this dynamo had acquired during the successful run of its patent, and from the published results of trials conducted at the South Foreland in experimental competition with other good machines. As the Gramme presents fewer constructional difficulties than some of its competitors, and produces a direct current at a moderate speed, it will be the best kind for an amateur to set about making, and presuming a fair knowledge of the use of tools, he will be able, in following these instructions, to construct a really useful dynamo after the style of Fig. 1, which represents one of the original machines.

CHAPTER III.

In starting upon the dynamo there will be first required an iron carcase, of such design as will produce a large electro-magnet, with its north and south poles resulting in the middle of the structure that is left uncovered with wire. An electro-magnet in its usual form consists of two bars of iron, having an end of each united by a cross-piece of the same metal, a quantity of insulated wire being wound upon the limbs or bars, so that when a current of electricity traverses the wire, it renders the free ends of the bars magnetic, the cross-piece remaining unmagnetised. In the small dynamo to be described, the cross pieces are formed by the standards, the limbs or bars will hereafter be known as the field cores, and are so wound with wire as to produce consequent poles of north and south polarity. By employing two bars, one projecting from each side of a pole piece, the winding of each bar must be in different directions; sufficient iron must also be used in the field cores, so that they do not become too soon saturated, and thereby prove detrimental to the steady working of the machine. Fig. 1 gives a general idea of the shape the dynamo is intended to assume when completed. It will, however, be observed in the illustration, that the brushes are held by bosses cast upon the pole pieces; whereas, in the dynamo to be described, this method will be dispensed with, and a more convenient arrangement employed by attaching a rocking holder to one of the standards of the machine. An increased portion of iron in these bosses at the polar extremities is not desirable, but rather a gradual increase of metal towards the crown of the arch. In the engraving, Fig. 1, lubricators are shown upon the top of each standard, in which there is a hole bored to provide a passage for oil to the bearings of the shaft; in the dynamo now to be described, however, its lubricators can be conveniently arranged in receptacles in the standards immediately over the bearings, where they can be snugly out of the way. An eye bolt also appears upon the upper pole piece, for the purpose of lifting or removing the dynamo by a hook or bar, and this may, if desired, be reproduced in the machine about to be commenced. Leaving these small preliminary differences, the necessary articles required for the complete machine may be summarised as follows:—

> The upper and lower iron field magnet cores and poles, Fig. 3 also shown in section (Fig. 2).
> A pair of iron standards (Figs. 4 and 5).
> Gun-metal bearings for the axle (Fig. 6).
> The steel axle (Fig. 7).
> Pair of gun-metal supports for armature (Fig. 8).
> Laminated punchings, for armature, R. (Figs. 2, 9, and 10).

Five screwed brass rods with nuts, for bolting punchings together.
The commutator, of copper or gun-metal segments, S.C. (Fig. 2); also (Figs. 11 and 12).
Supports for the brush rocker, in cast-iron, S. (Fig. 4).
The rocker, of malleable iron (Figs. 13 and 14).
Clamps for brushes, of gun-metal (C. Fig. 14).
Brushes of copper or brass, in wire or thin plates (Figs. 15 and 16).
Driving pulley, of cast-iron.
Two binding screws and connecting clips for wire ends (Figs. 17 and 25).
Insulating material, of sheet vulcanised fibre, and ebonite rod.
Lubricators of syphon form, in gun-metal.
Copper wire covered with cotton, for wrapping round field cores.
Similar wire, for armature coils.
Pair of brass bridge plates, B. (Fig. 2) in section.
Paint for ironwork, and varnish for wire.

Fig. 2 is a sectional view, illustrative of the arrangement of the field magnet cores F, their pole pieces G, and the rotating armature R, and commutator C within them.

The field cores are iron bars of round section, F, Fig. 2, and when they are wound with wire will become magnetic under the influence of electricity generated by the machine, and the iron cores thus magnetised will excite electricity in the wire coils of the moving armature R, by reason of the magnetism developed in two halves of its laminated iron core by the influence of the pole pieces attached to F, so that magnetism is produced in the field cores by passing round them the current generated in the armature, and they, by the inductive action of their pole pieces upon the iron body R of the laminated ring, develop electricity in the wire surrounding it.

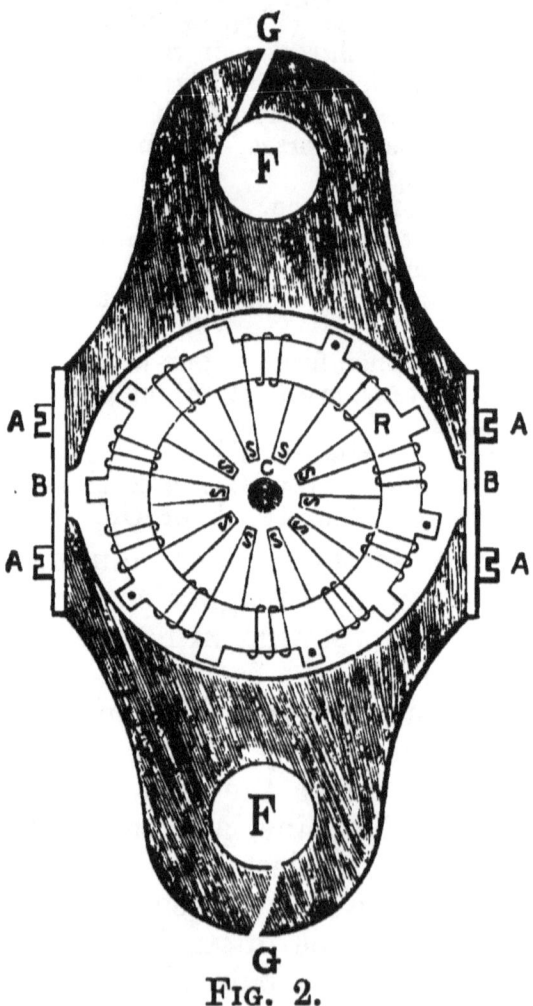

Fig. 2.

When the iron core, R, of the armature rotates, the magnetism arising in two halves of the ring does not travel with it, but remains stationary, and the wire coils covering the core develop their electricity by induction in their rapid passage over the *fixed magnetic regions of the ring;* it will be observed that the ring is wound with ten coils of wire, each of which is connected to a division of the commutator C : as the armature revolves the currents produced by the moving coils are picked up by the brushes near B B and conducted around the field magnet cores.

For the purpose of studying the principle of the armature, the ring core may be imagined as not being in motion, but the coils of wire revolving around it instead. Upon consideration it will be seen that the iron ring without rotation becomes magnetic when the pole pieces are magnetised by the fields being excited by a galvanic battery, or when they are formed of permanent steel magnets as sometimes employed in the hand dynamos of French manufacture; the revolutions therefore to be considered are those of the coils enveloping the semi-circular magnets of the ring. A line, supposed to pass from B to B, is the neutral line, where the induced magnetism does not reside in the ring, as it diminishes towards this point; the armature wire may be regarded as being divided into two halves, each end of the same sign being united at this neutral line by those segments of the commutator, C, that are engaged with the brushes diametrically opposite each other when passing through the neutral spaces, B.B.

CHAPTER IV.

The action of the field magnets upon the armature being understood, it will now be necessary to make a selection from three kinds of iron that can be employed in their manufacture.

THE FIELD CORES.

Wrought-iron, although the best, is inadmissible for the purpose of the amateur, if expense is of any consideration. There is the labour of forging, drilling, and tapping the bar ends for bolts, or tooling them down to spindle ends for screwing to receive nuts; also a great task would be involved in tightly and accurately fitting the pole pieces to their respective field cores. Malleable iron castings may be noticed, but their great drawback is the alteration of shape they are subject to in the process of annealing; therefore, cast-iron, if of good soft quality (which can be had) will answer well for the purpose; they will only be found a trifle hard where the metal is smallest in the casting, probably caused by quicker cooling than in the more massive portions of the metal. This can, however, be remedied by making them red hot on a smith's hearth over night, and allowing them to cool down very slowly in the ashes of the expiring fire; the process, if repeated a few times, will greatly improve them.

The dynamo under notice, being intended for 120 candle power (incandescent lighting), the field cores and pole pieces being of soft cast-iron, require to be of the size and shape indicated by Fig. 3.

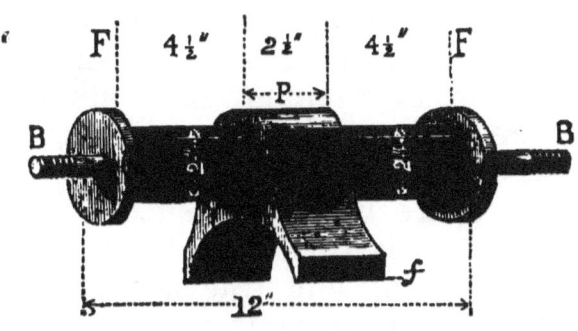

Fig. 3.

An advantage in using cast-iron is that the pole ends and field limbs are all in one piece, without joints; bad contacts of iron with iron in a built-up magnet cause a loss of magnetic power, so that any disadvantage of cast over wrought-iron is nearly counterbalanced in this respect in a small dynamo. A choice, however, may be made in the field cores, whether they shall have flanges of iron cast on their ends to form iron spools for the wire, or without these flanges, in order to slip separate coils of various gauges of wire on the field bars, for altering the purposes of the dynamo. Unless this may be required, it is advisable to have the flanges cast on the cores; the process of moulding them is somewhat difficult, but by the pattern being made in four removable portions it can be accomplished. In the centre of each flange the shank end of a wrought-iron bolt is embedded in the casting to provide a projecting pin, screwed with a $\frac{5}{8}$ in. Whitworth thread to secure it to the standard with a bright hexagon nut. A sloping recess, G, Fig. 2, should also be left in the sides of the poles, hollowed in the direction of winding for a stout and well insulated wire to lie in when beginning the convolutions over the field cores.

If possible, the field cores should be mounted in a lathe, in order to turn down the outside faces of the flanges to a bright and true surface, where they will be required to make contact with the standards. If the rim of each flange is also turned off bright it will greatly add to a neat appearance, and in cases of flanges not being employed the ends of the field bars should be turned off in a like manner, so as to ensure a good metal contact with the standard and be squarely bolted to it; a $\frac{1}{16}$ in. cut on each end of flange will bring them quite true. It is unnecessary to add that both the upper and lower fields should be of same length after being turned, which will make their distances between the standards 12 in., it being $11\frac{1}{2}$ in. within the flanges, as shewn in Fig. 3. One of the castings being selected for the upper bar, a hole can be drilled in the centre of the top of its pole piece for tapping to $\frac{5}{8}$ in. W.T., to receive a screwed eye bolt for lifting (if required) of 2 in. outside diameter. When the top and bottom fields with their poles in the centres are secured within the standards, the pole pieces should form a tunnel 5 in. and ⅛ in diameter, measured perpendicularly in the central portion of the chamber, which, however, is not quite circular, as the diameter should increase to 5¼ in. from an extremity of one pole piece diametrically opposite the other when they are finally adjusted and fixed; therefore, it will be observed that the iron lugs of the ring, and the wire wound upon it, are a little nearer the poles as they arrive within the crown of each arch. Nor do the magnetised poles entirely embrace the circumference of the armature; on each side a space is left between the polar extremities, and a central line imagined to pass through the centre of these spaces which divide the poles is termed the neutral line. Near to this line at the space inside B, is the position for the brushes to make contact with the sections of the commutator. When the poles are fixed in position, each neutral division between them should be spanned by a bridge of brass, B, Fig. 2, the plates of brass being secured by the screws, A; a flat surface provided at the extremities of the iron pole pieces, and shown at *f*, Fig. 3, is for the purpose of attaching an end of the bridge piece to it. This will serve three useful purposes; first to maintain the pole pieces in position, so as to prevent them from shifting and colliding with the armature when the latter is in rapid motion; secondly, since zinc, brass and copper are *insulators of magnetism*, these plates will intercept the lines of force passing between the extremities of the poles; therefore, the plates B, Fig. 2, may, with advantage, be made thicker between the ends of each pole piece, so as to fill the gaps on the inner sides of B. A final use for these brass bridges will be to engrave thereon the name of the maker; this, however, must be the last job in the construction of the amateur's dynamo.

CHAPTER V.

THE STANDARDS.

THESE are necessary for supporting the field magnets and to provide a connecting yoke for them, as well as for holding central brasses or bearings, in which the shaft may run. Fig. 4 is a sketch one quarter the full size of a standard, and a pair of these castings will be required. The illustration shows one intended to carry the brush arrangement, and a corresponding standard will be required for the other end—minus the brush support.

The circles marked C are bosses projecting from the inside surface about $\frac{8}{32}$ of an inch, and are intended to act as chipping pieces, to be filed truly flat, or, better still, removed by planing, so as to present a true and clean metal surface to receive the turned flanges, F, of the field cores (see Fig. 3). The holes in the centre require to be $\frac{5}{8}$ in. wide, and of slightly oval shape, so that a small adjustment may be given to the poles of the field magnets in fixing them over and below the armature. It is also as well to chip a countersink or bevel as shown on the edges of the holes; this allows the field cores to bed snugly home against the faced up standards at C, in the event of a burr being left on the inner end of the bolt in turning down the ends, or flanges, F, of the field cores.

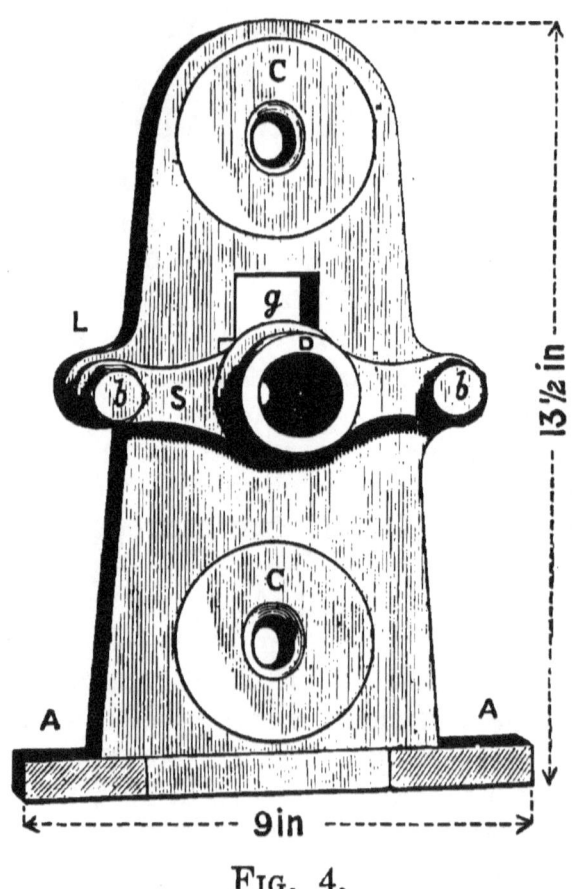

Fig. 4.

Referring to the pattern of the standard (Fig. 4)

L indicates a lug on each side; a ½ in. hole is left through each lug in casting, and to each hole is secured by means of ½ in. bolts and nuts $b\ b$ an iron bridge casting, S, having a hub, D, in its centre. This bridge can be easily mounted on the face plate of a lathe, in order that its cylindrical hub may be turned for the reception of a swivel brush rocker, which will be referred to hereafter. Holes ⅜ in. diameter should be drilled in the base of standards at A, for ultimately bolting the dynamo to the floor, —these holes are drilled more easily by inverting the casting for the operation; g is the receptacle for a lubricator, and the opening is arranged sufficiently wide to admit the brasses which fit into a space below. These split brasses may be observed in the cut behind the bridge S; they are secured in position by a malleable iron wedge, into which the lubricator is screwed by ⅜ in. brass gas thread, to be noticed presently.

The standard for the other end of the machine will be like that just described; but the lugs, L, in this case can be conveniently used as a very suitable position for the binding screws or terminals of the machine. The holes in the lugs must first be plugged with some insulating substance, such as vulcanised fibre or ebonite rod, and as the latter can be obtained of a suitable diameter for the purpose it may be preferred. When the holes are filled with this insulating material, it will require a $\frac{3}{16}$ in. hole drilled through each plug to receive the shank of a binding screw.

Having planed the circles, C, on the standards, and turned the ends of the iron cores of field magnets, as well as having trimmed the castings well down with an old file (removing all sharp edges and corners, &c., especially from the pole pieces), it will be advisable to frame them together in a temporary manner to obtain

THE IRON CARCASE

of the dynamo, of which Fig. 5 is an illustration

Fig. 5.

showing the field magnet cores (without flanges on them) secured to the standards. This is done by means of hexagon-shaped nuts screwed upon the wrought-iron bolts, B (Fig. 3), cast in the field cores, these bolts being supported in the holes which have been left at calculated distances from each other in moulding the standards. The pole pieces, P (Fig. 3), in the centre of the magnet cores require to be massive and smooth in the semi-circular sweep, and will, when one is placed over the other as shewn in the engraving (Fig. 2), be found by measurement to form a circular chamber within them of about five inches diameter. This can, however, be slightly adjusted by lowering the upper, or raising the lower field cores in order to bring the tunnel thus formed exactly to the radius of the shaft passing through its centre; when the armature core is in due course mounted on the shaft, and the true position of the pole pieces determined, they can be maintained in their proper places by some steadying pins projecting from C in the standards. Before this can be accomplished, however,

THE BEARINGS

to support the shaft will require consideration. In the standards (Fig. 4) it will be noticed that in the middle of each, spaces are provided for the gun metal bearings, Fig. 6 (sketched full size), to fit into; and over each space is a wider but narrow gap intended to receive a malleable iron plate casting tapering like a wedge, which, being tightly inserted in the slots over the bearings will firmly secure them. In the centre of the flat of this wedge plate a 1/16 in. hole should be drilled and tapped 3/8 in. brass gas thread, to hold the end of a syphon lubricator

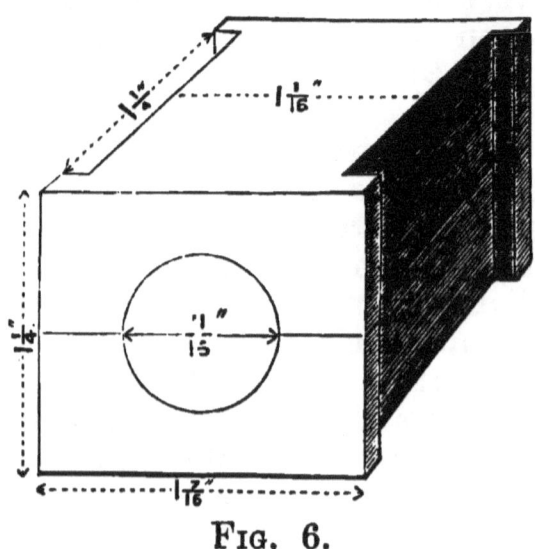

Fig. 6.

screwed to this size; the stem projects through the wedge plate and enters a hole provided in the top half of the bearings to conduct the oil as well as to make it impossible for the wedge to get loose. The chamber for the brasses will first require a little chipping to widen its narrowest side to the size of the opposite end (this tapering being a necessity in moulding) and will subsequently require filing to the width sketched within the flanges and shown on the top half of the gun metal bearings (Fig. 6).

The two corresponding halves of the bearings are to be filed to the sizes indicated on the sketch.

When they are nicely fitted in the standards, *without shake*, and the iron wedge tightly driven in over them, the faces may be filed bright, taking care to use a suitable file. The next consideration will be to enlarge the central hole, formed by the two semi-circular halves in the castings, to the size shown, viz., $\frac{11}{16}$ths; first, however, any grit or sand should be removed with an old round file, so that the tunnel is bright within and ready to receive the tapering end of a $\frac{5}{8}$ in. fluted rhymer, using a tap wrench with double arms for operating it, in the manner of tapping a thread, and working alternately from each end; a smooth and polished surface will now be developed within the hole, and when this size of rhymer has passed through, it should be followed by a $\frac{11}{16}$ parallel one, to be worked through in the same manner as the former tool. This will finish the bearings quite smooth, parallel, and correct; the split halves should then be numbered or marked, so that their positions may hereafter be known whenever they are removed from the standards.

CHAPTER VI.

Having finished the bearings in a workmanlike manner, the next job will be

THE STEEL SHAFT.

This must be selected from best mild round steel bar of one inch in diameter : the reduced ends of the bar to form the spindles can be tooled down by a smith, and this will save some labour and wear and tear of the lathe in removing the metal if the steel is sent to the forge ; but if the bar that is chosen happens to be perfectly straight it may be advisable not to let the smith manipulate it, lest he should upset its trueness. If not forged, the bar must be cut the full length required for the finished shaft sketched in Fig. 7, with the spindle ends and other dimensions illustrated.

If, however, the spindle ends e and $e\,f$ are tooled down by a blacksmith, the steel may be cut an inch or so shorter, as it will lengthen to the required size in forging the spindle ends for bearings to a smaller diameter. With an allowance of sufficient metal to remain for turning, by means of a pointed punch, lightly indent the middle of the end of the bar forging, that it may be mounted and twirled between the lathe centres to ascertain if it has been correctly

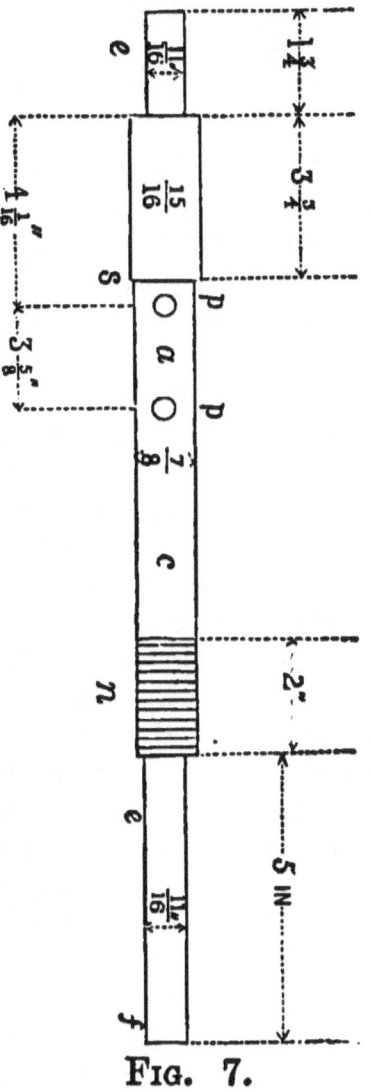

Fig. 7.

marked; if not, the experiment must be repeated

until the bar will revolve without wabbling. When this feat has been accomplished, the centres marked in the ends of the steel may then be deepened for turning, either by again using the pointed punch and a hammer, or by drilling, say a $\frac{2}{?}$ in. hole sufficiently deep to clear the extreme points of the lathe centres. It is important that the steel bar be perfectly straight and truly centred, because, at the end towards the shortest spindle only $\frac{1}{16}$ in. must be taken off in turning, as it is required to leave a diameter of full $\frac{11}{16}$ in. from the shoulder of the left hand spindle to the light shoulder at s. The space between s and the screw will be occupied by the armature supports at P.P., requiring $4\frac{1}{2}$ inches of the shaft, and the commutator $1\frac{3}{4}$ inches; their respective positions are denoted by the armature at a and the commutator at c. A suitable nut on n, with a washer intervening, will screw the armature and commutator snugly against s; a slit tube should be inserted between the armature supports to receive the thrust of the nut on n, the screw thread of which should, if possible, be chased in the lathe, of a pitch about twelve threads to the inch; $p\,p$ are $\frac{3}{16}$ in. steel clutch pins, tightly inserted in holes drilled into the shaft, and projecting about $\frac{1}{8}$ in. to engage the armature supports and prevent them from turning upon it.

It is perhaps needless to remark that the spindle ends e (see Fig. 7) must be finished perfectly smooth so as not to cut the gun metal bearings (Fig. 6); they must however, turn freely in them without shake. The extra length of spindle at *f* that projects beyond its bearing is to receive the driving pulley for revolving the shaft. This pulley may in size be adapted to the circumstances of the machinery intended to drive the dynamo; a small one, however, 2in. wide by 3½in. diameter, will be suitable for running from an engine fly wheel direct, or a rather larger one may be substituted where the motive power is obtained from countershafting running at high speed. The armature of the dynamo will be required to make 1,700 revolutions per minute to obtain the best results. The pulley casting should either be drilled out in the boss to $\frac{11}{24}$ths the size of the spindle, so as to fit tightly on the same without reducing it, but as it is difficult to select a drill that will produce a hole to the exact size required, it may be drilled somewhat smaller, say to ½in., and the bore enlarged in the lathe to the exact diameter required to obtain a tight fit upon the spindle. A keyway must now be cut in the boss of the pulley with a small square file, and a flat surface filed upon the end of the spindle as a seat for the steel wedge key that the pulley may be held fast upon the shaft for turning to a true and bright surface. There is a disadvantage however, in turning up a pulley on a slender shaft, and a short mandrel may be employed for mounting it to be turned in the lathe. When transferred to its intended quarters on the shaft a little wabbling is apt to be perceptible; this will certainly occur when keyed on if it has been allowed to fit loosely on either the mandrel or shaft in turning.

CHAPTER VII.

Presuming the steel shaft and its iron pulley to be finished in compliance with the foregoing suggestions, the next matter will be to fit

THE ARMATURE SUPPORTS

upon the shaft, and but little consideration is required to perceive that iron would be inadmissible for them, therefore they must be formed of brass or gun metal. Their shape or pattern, however, varies with different machines and makers, some preferring them in

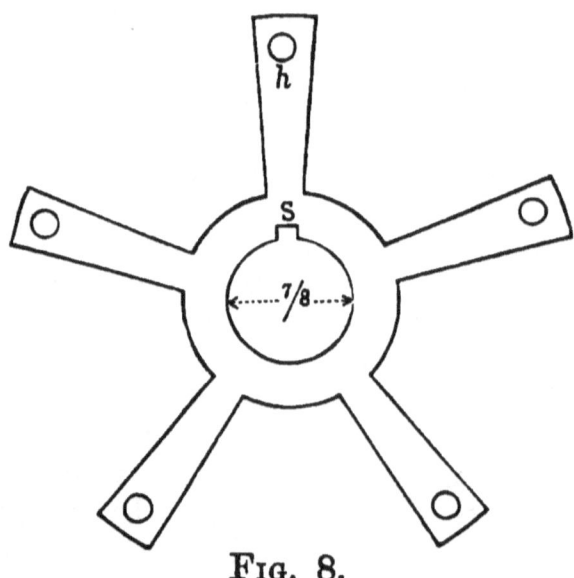

Fig. 8.

the form of wheels, while others use spiders or star-shaped flanges as shewn in Fig. 8, the sketch being a half-size representation of the supports required for the amateur's dynamo. A pair of castings will be necessary; they can be obtained with a hole cored in the centre of the boss in casting, the diameter of which requires to be $1\frac{1}{2}$in., with five projecting arms (to suit the toothed-ring core they will carry), by corresponding to alternate cogs from which $\frac{1}{4}$in. brass rods extend to enter the holes h; the boss has an extra length on one side to obtain a firm support upon the shaft; the hole left in the centre of the casting can be bored in the lathe to a sliding fit of the size shewn on sketch, which it will be observed, is the diameter of the shaft at a to be occupied by the armature.

The best method of mounting or chucking the supports in the lathe is to fit a bit of boxwood in a cup, or other convenient chuck, and hollow it out, so that the respective bosses of the stars may be inserted tightly in the boring of the chuck; the hole in each support can then be enlarged by a slide rest cutter, to obtain an easy fit upon the shaft at $p\ p$ (see Fig. 7); while mounted in the lathe a pointed scriber is adjusted in the slide rest, and with a little manipulation of the gut band, by hand, a radial line can be marked on each arm, to coincide with the centres of the holes in alternate cogs of

THE ARMATURE CORE PUNCHINGS,

Fig. 9, the holes h in the arms of the star supports

being drilled at the marked line to a ¼ of an inch in diameter. The groove or keyway *S* in each boss must be filed out to receive the projecting pins *p p* on the shaft, and these slots must

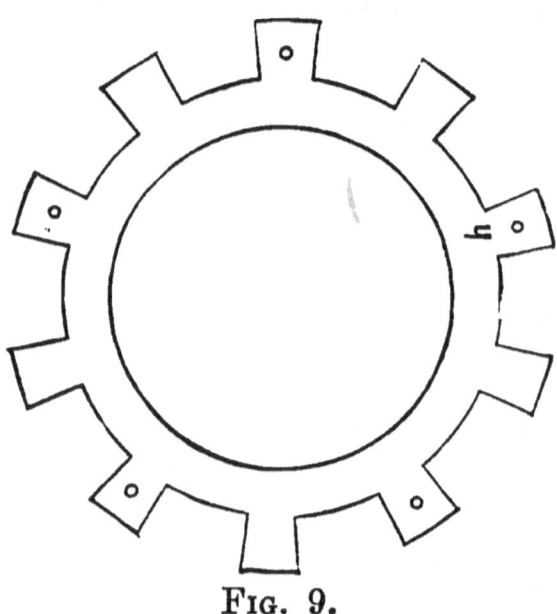

Fig. 9.

be coincident in each support, the catch pins *p p* being inserted directly in front of each other. The bosses, extending on one side of the star supports when mounted upon the shaft, are to face inwards towards each other; the holes *h* in the arms should easily admit the projecting ends of the screwed rods passing through the holes *h* of the punchings (Fig. 9) which bind them together; in Fig. 10 small hexagon brass nuts are illustrated on alternate cogs, showing the method of compressing the lamination plates of soft sheet iron together; these punchings can be obtained from electrical instrument dealers, of five inches external diameter, to suit the chamber formed by the semi-circular pole sweeps; if the punchings are thick about sixty of these iron stampings will be wanted for their combined thickness to equal the width of the pole pieces; if thin, a

Fig. 10.

greater number will be wanted. Before bolting them together they should each receive a coat of special flexible varnish, by means of which tissue paper can be attached to one side of each punching.

As the holes for bolting are alternate on the ten lugs of the stampings, five brass rods of ¼ in. diameter will be necessary to build up the laminated core, and this process will be simplified by having the rods a little longer at first, and shortening their length after the laminations have been compressed into a thickness of 2½ inches, using ¼ in. brass hexagon shaped nuts on the screwed ends of the rods outside the core. These are gradually screwed up evenly all round the ring, and when the screwing is completed the nuts may be permanently secured to the brass rods with a touch of solder, using powdered resin as a flux.

The depth of each space for filling with wire between the cogs being half an inch, the wire convolutions will consequently overlap equal to the depth of winding, on each side of the laminated core; therefore for the supports to clear the ends of armature wire coils they must be held at ⅝in. distance from the ring core by using "set-off pieces" of small brass tube ½in. long, to be slipped upon the screw of the brass rods outside each nut securing the stampings, and inside the arms of the star supports.

In case of an arm of the star not presenting itself fairly to the end of a screwed rod proceeding from a lug of the core, a smart tap administered with a light hammer to the arm, or with a mallet to a rod if it is bent, will soon remove the difficulty.

CHAPTER VIII.

The core may now be mounted upon the shaft to see how it runs with regard to trueness, and if the holes in both supports have been carefully marked and drilled, and the screwed rods holding the core are not bent, it may be expected to run accurately; any little eccentricity of behaviour can, however, be corrected by changing the screwed rods to different arms; if necessary for clearing the pole pieces, the edges of the cogs may be turned off a trifle by employing the lathe back gear for the purpose. The fastening nuts should, however, be first screwed on the rods, outside the arms of each support, and the superfluous ends of the screws projecting beyond the exterior nuts removed by a hack saw, leaving a bare sixteenth to form a burr to prevent the nuts from working off. It will be obvious that before turning the exterior of the core, it should be firmly secured in its position by slipping washers over the shaft at c and tightening all up by the nut on screw n. (See Fig. 7.)

When this built-up core is mounted upon the shaft and found to run truly between the lathe centres, it should be marked so as to denote the arms of the supports fitted to certain ends of the brass screwed rods projecting from intermediate cogs on the ring core, in order to ensure its being again mounted in the same position after it has been removed for the purpose of winding; the ends of the brass rods must not, however, be burred upon the nuts until after the wound and tested ring is mounted upon the shaft. To form a burr on the rod ends, the armature should be held vertically, the lower end of the rod being placed upon some solid substance while the upper end is tapped with a hammer.

The next operation will be to trim up the core, using a half-inch square file to remove any little roughness on the sides of the cogs produced by the edges of the laminations,—a little time spent with a file will bring the compressed mass quite smooth and solid in appearance; the interior of the ring should also receive similar attention, using a rather large half-round file; and having so far prepared the laminated ring core for insulation, the next matter to be put up in hand will be the commutator.

It is a subject for consideration whether the device known as a *commutator* in the Gramme dynamo might not as well be regarded as a *collector*; so long, however, as the amateur recognises its purpose, the mere term by which this arrangement is named becomes unimportant. It consists essentially of a series of copper, gun metal, or phosphor bronze bars, arranged parallel with one another, in equal divisions of a circle, and corresponding in number to the coils of wire upon the armature ring. Each segment of metal is insulated from its neighbour. Upon reference to Fig. 9, it will be observed there are ten spaces between the cogs of the laminated core intended to receive wire, consequently the same number of divisions will be necessary in the commutator; to make one for the small dynamo now progressing, there will be required a gun metal cylinder casting, 1¾in. wide, 2$\frac{7}{16}$in. internal, and 2$\frac{15}{16}$in. external diameter, leaving a shell thickness of full ¼in. A piece of good sound boxwood must now be selected to form a hub upon which the cylinder has to be mounted, after its interior has been made tolerably smooth by a light boring cut in the lathe. This hub must also be bored to ⅞in. diameter to fit tightly upon the shaft at C, Fig. 7, and is then turned up in order to receive the cylinder casting, which must fit tightly over the hub. Fig. 11 is a full size sectional representation of the boxwood support H, with the gun metal cylinder W upon it. Hard vulcanised fibre is also a good material for forming the hub.

The mounted cylinder should now have a light cut taken off the ends in the lathe, and its periphery rubbed bright with an old file while revolving in the lathe, when it will be ready for dividing into ten equal sections to correspond with the recesses in the ring Figs. 9, 10, and unless the amateur is the proprietor of or has access to a lathe with a division plate attached to it, this task of marking had better be left for an engine divider to set twenty division lines equi-distant from each other around it. Ten of these lines may be cut deep, as they have to be sawn through with a metal cutting saw, the other intermediate ten need only to be slightly marked, as they are simply for denoting the position for a hole to be drilled through the centre of the ends of each metal segment into the wood within to receive a ⅜in. brass screw S, Fig. 11, for holding the segments of metal upon the hub. These screws should enter the boxwood rather tightly, and the heads be sunk

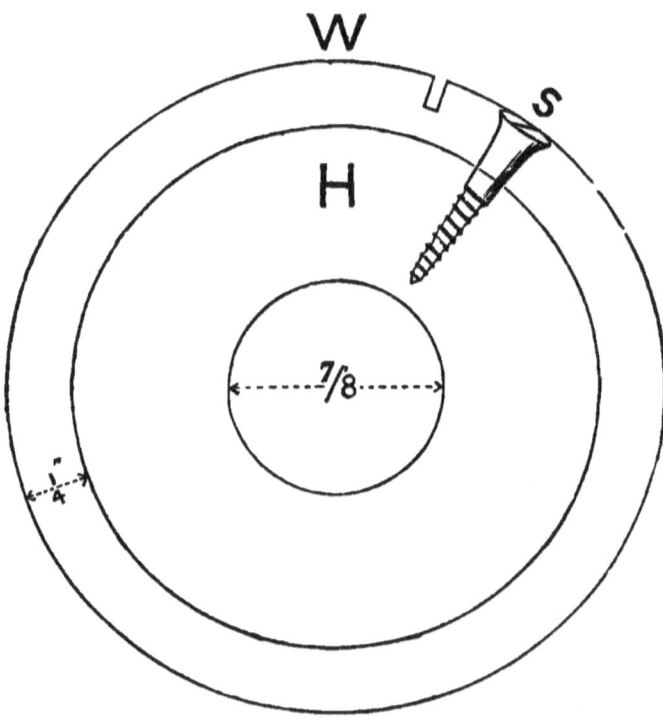

Fig. 11.

flush in the metal by a countersink, care being taken that they do not reach the shaft. If, however, the screws do not exceed ⅜in. in length this will not occur. There being ten segments to be formed, twenty screws will be required, and when they are inserted, a sharp hack saw applied to the deep lines between the screws will neatly and truly divide the cylinder into ten parts. The commencement of the saw cut may be observed between W and S, Fig. 11; the saw should also enter the boxwood to a depth of ¼in. for the insulating medium to obtain a firm support, and prior to the complete

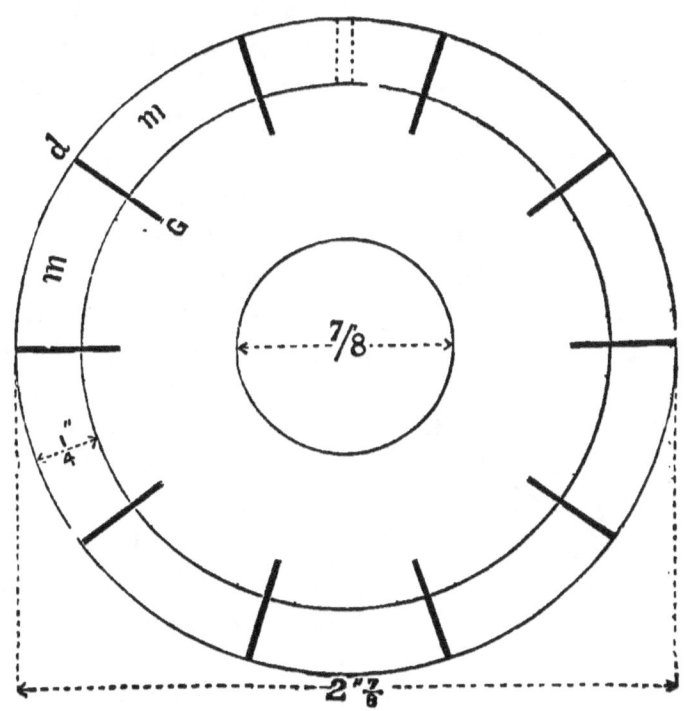

Fig. 12.

separation of the cylinder into segments, it can, if necessary, for trueness, be mounted in the lathe to receive a light cut on its periphery by the slide rest tool, but generally speaking, the application of a file to the surface of the cylinder when it is revolving in the lathe will remove any rough edges that may exist.

The ten metal segments must now be insulated from each other by slips of hard black or red vulcanised fibre being tightly fitted and glued between them in the saw cuts, see Fig. 12 d—G, shewing sectional view of the fibre slips insulating the neighbouring divisions each from the other.

The insulating substance should be of sufficient thickness to require pressure to insert it between the sawn grooves, and any protuberance of this material beyond the metal divisions can be easily pared off flush with a sharp penknife or carpenter's chisel; two hard black fibre discs are now required to form cheek plates for the sides of the commutator, as security from the segments flying off; these discs should be made from sheet fibre $\frac{1}{4}$in. thick, and a square piece of it $3\frac{1}{2}$in. × $3\frac{1}{2}$in. can be mounted upon a face plate in the lathe by a screw at each corner of the square. Since it is easy material for turning, a circular recess $\frac{1}{8}$in. deep can be readily sunk (using a wide wood-turning or carpenter's chisel) to the exact diameter of the outer circle of metal segments, viz.: $2\frac{7}{8}$in. diameter, leaving the rim, margin, or flange so as to overlap and fit tightly around the metal divisions. A depth of $\frac{1}{8}$in. will be sufficient to project beyond the segments as a rim; a corresponding plate must also be made for the other side of the commutator, and both of them can, if desired, be fastened to the metal segments or boxwood hub by screws; the nut n upon the shaft will, however, be sufficient to secure the rims upon the segments, thus confining the ends within the sunk circle of the vulcanised fibre discs, so as to prevent the segments from working off or getting loose when running at high speed.

It will now be necessary to provide a conductor to electrically connect each segment of the commutator with two adjacent coils of the armature; for this purpose drill a sixteenth hole near the end of each division of metal intended to be placed towards the armature, shewn in section by dotted line through the top segment of Fig. 12; each hole being required to admit the end of a short length of No. 16 brass wire. The segments can be removed one at a time in order to countersink the under part of the hole, and when the wire is inserted it can be permanently fixed with a drop of soft solder in the countersink; the free end of each wire proceeding outwards from a segment should be tinned with solder in readiness for its being ultimately connected as a junction conductor with the union where the ending of one coil joins the beginning of the next one upon the armature ring core.

Having completed the commutator it can be mounted upon the shaft with the armature core, in order to ascertain if a washer is required between the boxwood hub and the armature support; it is better to use two thin lock nuts for screwing all up securely upon the shaft, as they are less liable to work loose than a single deep one.

CHAPTER IX.

The armature core, held by its supports upon the shaft, should now be mounted in the bearings of the dynamo carcase, and by adjusting both field bars in each hole of standards, the pole pieces can be closely and accurately set to clear the teeth of the ring. The nearer the latter can revolve without colliding with either pole piece the better for inductive effect, and about $\frac{1}{24}$in. clearance may be allowed. When the exact position of each pole extension has been determined upon, two $\frac{1}{4}$in. holes must be drilled in each standard, one under the hexagon nut, the drill passing into the flanges of field magnet bars, *but not through them*. The object of these four holes is to receive steady pins of $\frac{1}{4}$in. round iron rod, slightly tapered and hammered permanently into the standards, leaving $\frac{3}{16}$in. projecting from the planed surface inside, in order to fit into the holes in flanges to retain the field cores and poles in proper position, and so that they may also readily fit together again correctly, when the machine is at any time required to be taken to pieces.

The brass bridge plates shewn at B, Fig. 2, will also materially assist in maintaining both pole pieces in their required position; and they may now be fitted upon the extremities of the pole pieces with small iron or brass screws shewn at A, Fig. 2; previously, however, measuring each gap between the pole ends by a pair of inside callipers. The spaces will be about 1⅜in., and the pole extensions which may happen on one side to approach each other more closely than the opposite ones may be corrected with a file so as to equalise the distance of both neutral spaces, if any difference should be found to exist between the opposite gaps.

THE BRUSH ROCKER.

By preference this should be a malleable iron casting of the design illustrated by Fig. 13, and R

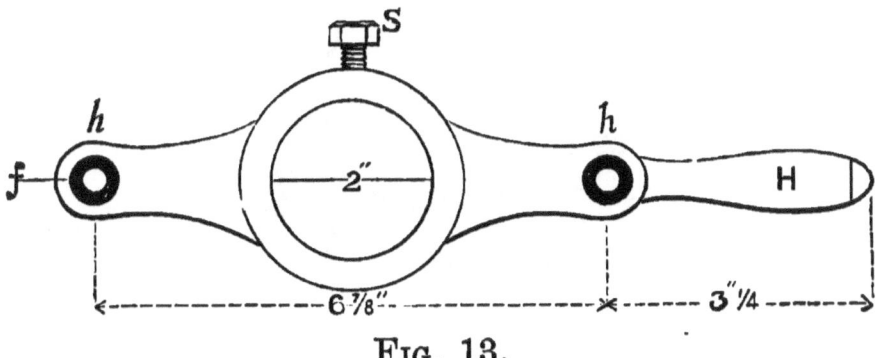

Fig. 13.

in Fig. 14, which can be easily bolted on to a face plate by the ½in. holes h left in the arms in casting; the enlarged centre can then be bored to the diameter shewn, that it may be set at any desired angle upon the turned hub D of the support S on the standard, Fig. 4, in order to adjust the brushes upon the commutator; a set screw S is used for retaining the rocker in any desired position when the brushes are properly adjusted. H is an ebony handle held upon

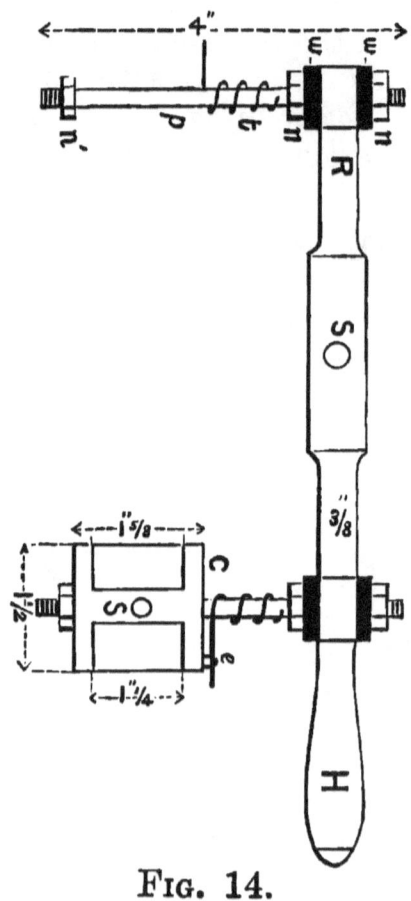

Fig. 14.

a ¼in. rod projecting from one end of the rocker; a

second handle on the opposite end can be added if desired. The holes h require to be plugged with $\frac{1}{2}$in. round ebonite rod; and a $\frac{1}{4}$in. hole shown at f, is afterwards drilled through this insulating substance.

In Fig. 14, the rocker R is illustrated with one of the gun metal brush clamps C attached to, but insulated from it: the clamps are supported on spindles p and have a free vibratory movement upon them. A $\frac{1}{4}$in. hole is drilled through a boss on the casting underneath the receptacle for the brush and parallel with the bridge S carrying the set screw; these spindles can be conveniently made from $\frac{1}{4}$in. round brass rod of the length shewn on sketch, being secured to the brush rocker R by nuts $n\,n$, and insulated from it by the vulcanised fibre washers $w\,w$ $\frac{1}{8}$in. thick and 1in. diameter, and the plug of ebonite f in Fig. 13. A nut n' on the end of each spindle will prevent the clamps from working off, and an additional nut will afford connection for a wire leading to a terminal if shunt wound; a tension spring t having one end of its spiral fixed to the pin p by passing through a hole drilled in the stem, and the other end entering the loop e on the side of the clamp, can be adjusted to exercise a light pressure of the brush upon the commutator as the brush clamp is tilted by the action of the spring.

THE BRUSHES.

These can be made as illustrated in Fig. 15, of brass or copper wire of No. 20 gauge, cut into a number of six inch lengths; one end of each must be separately tinned with solder, then packed in a

Fig. 15.

temporary rectangular shaped tin tube, with the tinned end projecting from it about one inch for binding them together by soldering, the interior of the tubular mould to be equal to the inside of brush clamps. Then the bundle can be removed from the mould, and a narrow band of copper or tin substituted so as to slip somewhat tightly over the wires towards the free ends and confine them closely together. Very good brushes can be more easily made from thin brass plate rendered springy by hammering, using two or more plates in layers one over the other, each having a series of slots or fine saw-cuts as represented in Fig. 16.

Fig. 16.

These can be set as required in the clamps and held fast in them by a screw S in the centre of the bridge. The brushes must be arranged not to press *heavily*, but *evenly* upon the commutator; and jumping must be guarded against to prevent sparking.

The neutral points will not be found between the extremities of the pole pieces; but somewhat in advance in the direction of rotation, and are variable according to the speed of armature, as well as from any alteration of resistance in the lamp circuit. A "lead" has therefore to be given the brushes which require adjustment upon an increase or decrease of the current generated in the armature. A little end-play of the shaft is sometimes allowed, in order that the brushes may not wear the commutator unevenly, and to prevent heating.

TERMINALS

Are now required to connect together the dynamo wire circuit with the outer wires running to the lamps; one standard has the brush arrangement bolted to it, the opposite one is intended for the terminals to be attached to the lugs on its sides. The holes in them are plugged with ebonite rod and a $\frac{3}{16}$th hole drilled through each to receive a pair of

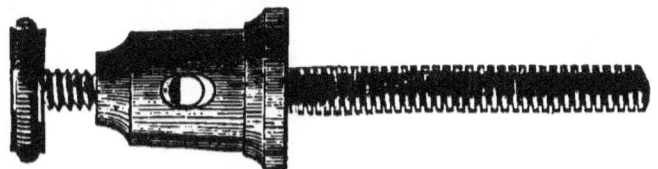

Fig. 17.

substantial terminals, with extra long shanks (as illustrated full size in Fig. 17) and stems sufficiently long to pass through the lugs and receive a nut at the end of each. This will not only secure them to the standard, but also provide attachments for the wire of the dynamo circuit. Insulating washers similar to those used upon the pins of the brush rocker are required on each side of the lugs. As a pair of ready-made terminals, like those shown in the wood-cut can be bought for 1s. 6d. with nuts, it is hardly worth while for an amateur to make them; should he, however, wish to do so the exact size of a suitable pattern is before him.

CHAPTER X.

The fitting up of the dynamo being now finished, a preliminary trial can be made of its mechanical qualities by attaching the band from a lathe flywheel to its pulley, and presuming the armature core to rotate freely within the chamber formed by the pole pieces without touching, with the brushes slightly pressing upon the segments of the revolving commutator, and the shaft turning in a satisfactory manner, the aspiring amateur, thus encouraged, will enter upon the second stage in the development of the small dynamo, by insulating the field magnet cores and the toothed armature ring preparatory to winding.

The field bars should first receive a coating of Japan black, or varnish paint, such as Aspinall's enameline, which is not only a step in the direction of insulating the wire from the iron, but will prevent rust, which might cause mischief at a subsequent period. The writer on unwinding a number of electro magnets belonging to Morse telegraph instruments has frequently found the insulation of the innermost coils of wire destroyed by rust on the iron core, probably caused by the moisture of an operator's hand in winding. The circumstance is noted as a hint to those who would wish to secure their work from such danger; it may also be remarked that malleable iron is very susceptible to rust, especially by the action of methylated spirit used in shellac varnish.

After the paint has dried upon the field cores, they should be covered with broad white cotton tape (to be obtained at any draper's), securing the ends of wrapping with a touch of joiner's glue. Eight thin vulcanised fibre or india-rubber collars must now be cut from a sheet of either material to insulate the inner surface of iron flanges on field cores, also the sides of pole pieces; to fit these collars over the field bars they will require to be cut in halves, or clipped in the manner of messengers (as used by boys when flying kites). They must be attached to the iron with thin glue, Giant cement, or Kay's coaguline; and the tape should be well soaked with melted paraffin wax.

Certain teeth of the armature core being marked to correspond with definite arms of the star supports, the laminated ring may be disconnected and painted or japanned prior to insulating it with tape. The sides of the recesses must also have strips of tape cemented to them, and all portions of the ring to be occupied with wire require to be covered with the same material, which should be finally basted with melted paraffin wax. Too much care cannot be exercised in the matter of insulating the iron ring.

SERIES, SHUNT, AND COMPOUND WINDING.

At this stage the amateur will have decided whether the dynamo is to be series, shunt, or compound wound. In the series arrangement the field magnet cores require to be wound with wire of a large gauge and of comparatively few turns, while those in the shunt system are wrapped with a greater number of convolutions of smaller wire. In the arrangement of the latter only a portion of the current generated by the armature is used in the fields, whereas, in the series-wound dynamo, the whole of the current produced by the armature is passed through the fewer turns of wire upon the field cores; the same magnetising effect occurs, however, to the iron in either case—for example, field magnets as series wound require a certain number of comparatively thick turns of wire, and the shunt a great many more turns of finer wire, for on a field magnet core of given size one ampère can be caused to magnetise it as fully as 1000 ampères in the following manner.

Supposing the experimenter is only employing one ampère, then he must increase the number of ampère turns of wire; therefore, if he uses 1000 turns of a fine gauge, there will be a magnetising effect as powerful with one ampère, as there would be with the 1000 ampères passing once round the iron core. It takes a requisite number of ampère turns to magnetise to saturation the iron core of an electro-magnet, and it is immaterial whether a large current traverses a few convolutions around the bar, or a small current circulates many times around a large number of turns of wire wound over the same bar. This winding with fine wire explains the principle of a shunt machine where the winding has a higher resistance in the fields which only receive part of the current generated in the armature, the other portion supplying the leads to lamps; in the case of incandescent lamps being added to the main circuit the shunt dynamo can meet the extra demand upon it, as the additional resistance to the main circuit diverts the current into the shunt coils, and consequently increases the power of the magnetic field; upon the removal or extinction of any lamps the opposite effect occurs.

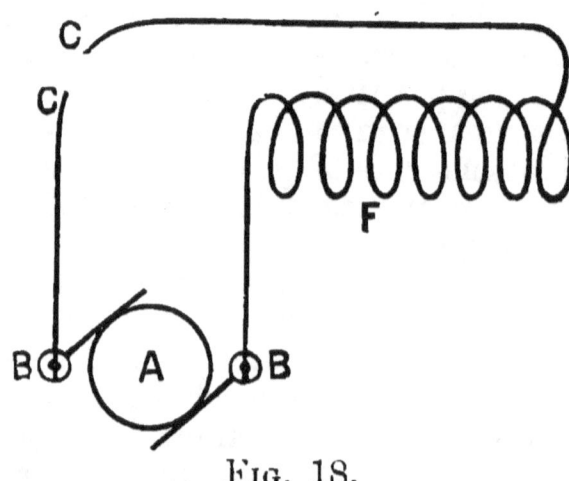

FIG. 18.

Fig. 18 illustrates the circuit of a series dynamo

adapted for arc lighting by means of two carbon electrodes C. C.; incandescent lamps, however, require the E. M. F. to be steady, as they are apt to give way if an excess of current is supplied to them; the shunt-wound dynamo gets over this difficulty by checking any rise of current consequent upon a variation of resistance in the lamp circuit, for as the resistance is lessened by turning off lamps, so a reduced portion of current is served to the electro-magnet coils F.; this consequently weakens their action upon the armature. It will be seen from the diagram, Fig. 19, which indicates the main and shunt circuits of a shunt-wound dynamo the current generated by the revolving armature A is split or divided at the brushes B between F and L; F denoting the field magnet or shunt circuit, and L the lamps in main circuit.

Referring to the series-wound dynamo in the diagram Fig. 18, it will be observed that the work to be done in lighting is included in one circuit, viz., the revolving armature A, the field magnet coils of thick wire P, and the lamps. In a parallel arc arrangement, each lamp acts as a bridge between two branch wires proceeding from the terminals.

The shunt-wound system shown in diagram Fig. 19, is better suited for an installation of incandescent lamps, in consequence of the power of this form of dynamo being greater or less as may be required, the work to be done in the external circuit L of a variable number of lamps in use being regulated by an even balancing of supply and demand.

The field magnet, armature and lamps, all acting harmoniously together in this system of adjustment.

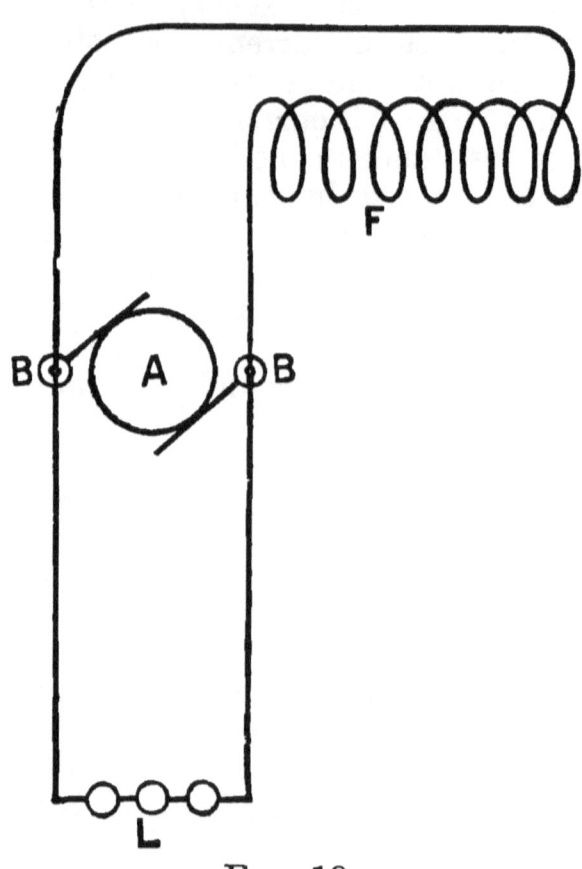

Fig. 19.

In the series machine, there is but one path for the current through the field magnet and armature coils, conductors and lamps; whereas in the shunt dynamo, there are two circuits to be considered, viz., the *main* and the *shunt;* in this arrangement the current divides itself between the leads, lamps, and shunt coils wound over the iron cores of the field magnet.

In the other method of winding known as the compound or series-shunt system (which, however, is not advised for small dynamos) the generator is rendered automatically self-regulating, so that it can supply the current in proportion to any number of lamps in use, the electro motive force remaining constant. The method consists of double winding, that is to say, series and shunt combined; the windings may be arranged in various ways, but in general practice the series wire is first wound over the field bars, and afterwards the finer shunt wire. In the case of more lamps being put on in parallel form the external resistance is decreased, and the increased current round the series coils produces an increase of magnetism proportional to the requirements of the case.

A portion of the electricity developed in the series-shunt wound dynamo is employed to maintain the magnetism of the field cores and pole pieces, the remainder being used for the lamp circuit; the magnetic field increases and decreases, as the resistance in the external circuit varies. The shunt wire is often termed a "teaser;" Brush on the other side of the Atlantic, and Varley on this, both claim to have originated the idea of compound winding in the series-shunt system; Varley also suggests winding two of the four-field magnet cores with thick and two with small wire, using the former as series and the latter as shunt circuits; or, in another method he proposes winding one end of the core with thick, and the other with thin wire, which, it is suggested, will answer as well or better than the usual compound system of winding, as wire can easily be added or removed in coils so as to get the exact adjustment.

CHAPTER XI.

If the dynamo is to be wound for arc lighting, or for a definite number of lamps, not subject to variation, wire of a large gauge only will be required for the fields. No. 16 covered with two layers of cotton will be suitable, and the quantity must be regulated by the amount and resistance of the armature coils.

Professor Sylvanus Thompson asserts that to determine the gauge of wire to be used for the field magnets of series wound machines, *it is a useful rule to remember that the resistance of the field magnet coils of a series dynamo should be a little less than that of the armature, say two-thirds as great.*

In armatures of the ring type, as in the Gramme, only one quarter of the entire quantity of wire upon the core should be considered, because when the armature is running its wire is divided into two portions at the sections of the commutator, diametrically opposite each other at the brushes. The theoretical problem of winding a dynamo is, however, not clearly defined to the professional electrician, therefore the amateur might well be pardoned for getting confused in calculating resistances of dynamo wires. These are affected in various ways, such as increase of temperature when at work, also quality of the copper conductor, as well as the degree of softness of the iron employed in building up the armature core, which, if of hard nature would cause some heating, consequently affecting the resistance of the wire around it.

Large shunt dynamos have the field cores wound with a considerable resistance in comparison to the armature wire. This difference of resistance does not apply in the case of winding a small machine intended to light but a few lamps; as an instance, a small 100 candle power Gramme has been wound to act either as shunt or series; for the latter, by reducing the resistance of its field wire by coupling the coils in parallel system, the whole resistance of the wire circuit of the four limbs of magnet being employed when used as a shunt machine. In this experiment the field wire was 16 gauge, and the armature coils 18, but for continuous working under these conditions as shunt some heating of the armature coils might ensue.

Small shunt wound dynamos for lighting only six or eight 45 volt. lamps should contain a field resistance of not less than double, and not much over treble the entire resistance of armature coils: within these limits, the same size wire such as 18 or 20 may be used throughout. A frequent error in many small shunt machines in which the fields are wound with too high resistance, may often be remedied by removing wire, substituting larger, or connecting the coils in multiple arc arrangement.

If series wound the armature coils require to be of a gauge four sizes smaller than the fields. Again, the rules which apply to large dynamos are not deducible for small ones employing cast-iron for fields, therefore the sizes and approximate quantities of wire which have been found by experiment well suited to the small type of dynamo now ready for wiring are as follows:—

Field Magnet.	Armature.
Series wound...... 11lbs. No. 16	4lbs. No. 20
Shunt wound...... 12lbs. No. 18	$3\frac{3}{4}$lbs. No. 18
Compound wound $\begin{cases} 3\text{lbs. No. 16 as series} \\ 9\text{lbs. No. 20 as shunt} \end{cases}$	$3\frac{3}{4}$lbs. No. 18

It is important to select the best quality of copper conductor; that there is a vast difference in copper wire may be gathered from the remarks of the President of the British Association, September, 1888, with reference to the impurity of some copper conductors. Allusion was then made to the conductivity being in some cases reduced by as much as 50 per cent.

The copper conductor should be evenly covered with two layers of cotton, free from kinks; second-hand wire must be rejected; it should also be procured on bobbins or drums, or transferred to them before winding. These should have a spindle (a stair rod does nicely) passed loosely through the hole in the centre of each, and the winder will not then experience the difficulty that might happen in using it from hanks, with the consequent risk of entanglement, and probable loss of temper.

WINDING THE ARMATURE.

Before proceeding to wind the iron ring core of the armature, four slips of wood, each four inches long, are wanted; two of these pieces in section G, Fig. 20, are required to fit between the outside lugs L of the ring to support by means of the screws S

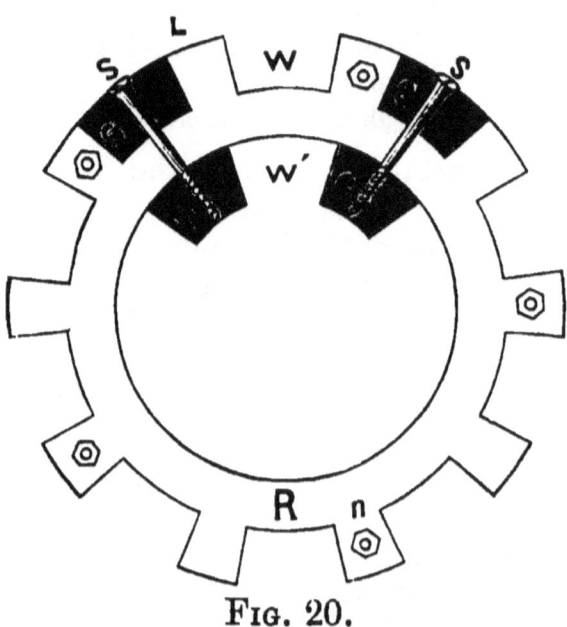

Fig. 20.

the two other slips at guide blocks inside the ring. Temporary walls are thus formed to limit the space W' in width for winding the interior of the laminated core. Without these guides the inner winding of the first coil would trespass upon the space required for the neighbouring one.

When the first coil has been wound it will be obvious that only one guide will be afterwards needed; the sides of a coil acting as a barrier, until the last recess has to be filled with wire, when the other guide can also be dispensed with; the strands on both sides will then leave the exact space between them for winding the final coil over the ring.

To wind the armature core it will be necessary to make a wooden shuttle as represented by Fig. 21,

Fig. 21.

nine inches long, and one inch full width, having a gap at each end $\frac{3}{4}$ in. deep by $\frac{5}{8}$ in. wide; its thickness may be about $\frac{1}{8}$ in. and the edges must be rubbed quite smooth with glass paper. Presuming the dynamo is to be shunt wound, fifty feet of No. 18 double cotton-covered wire must be cut off and wound upon it lengthwise; this quantity can be nearly all used for a coil with careful manipulation, making for the ten coils about 160 yards, or $3\frac{3}{4}$ lbs. on the entire ring.

The services of another votary of science will

be useful to receive the shuttle as it is passed through the interior of the coil, the winder, who must carefully guide the strands evenly and tightly in the channels between the edges on the periphery of the core, keeping them well within the guides inside the ring by piling the strands over the layers. The winder should occupy a low seat with his assistant on his left, and have at hand some small pieces of tape in readiness to protect the covering of the wire when taking it past any edges of the cogs that may appear dangerous to insulation; a hammer and a piece of wood hollowed concave like the outside slips holding the guides in Fig. 20 will also be required. The wood is for the purpose of being held upon each layer of wire as it is laid in the recess and tapped down flat with a hammer as the winding proceeds, so as to keep the coils from bulging at the exterior of the ring beyond the cogs, the uniformity of the strands lying *within the core* being of no consequence.

By means of a galvanometer and battery, tests for insulation should be frequently made by connecting with copper wire one terminal of the battery to one binding screw of the galvanometer; the other battery terminal leading to a free end of the wire coil being wound, another wire proceeding from the other binding screw of galvanometer and placed in contact with the iron core, will deflect the detector needle if the insulation is defective; if no deflection occurs the insulation is good and the winding may be proceeded with.

When a coil is wound a few spare inches of wire

should be left and bent round the end of a guide and the strands afterwards painted with shellac varnish. When dry a second and third coat can be applied so that the wire will become firmly set, then the guides can be removed and the sides of the strands of wire within the ring can be also varnished. This will set the inner layers fast together and secure them in their allotted space, the varnished strands forming a boundary wall to wind the adjacent coil against.

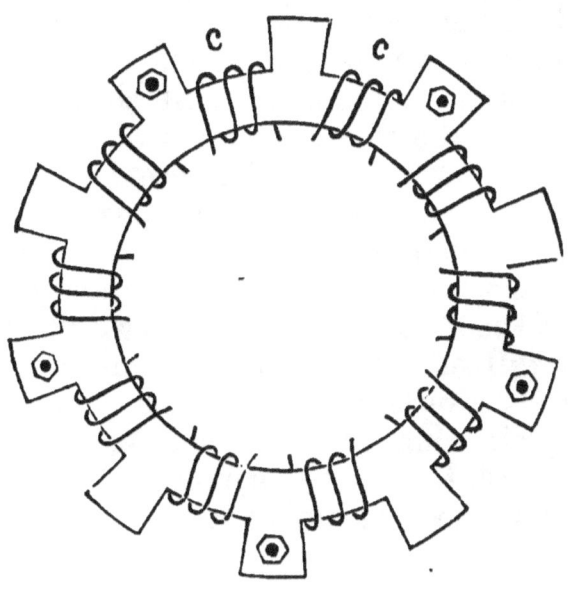

FIG. 22.

The first coil being wound as represented at *c* on the left Fig. 22, the next one *c* on the right must also be wound in the same direction, and the remainder of sections on the ring filled in the same manner; all beginnings as well as endings of coils require to be on the same side of the ring to connect them to the bars of the commutator. It is advisable to paint all the beginnings of coils with red sealing wax varnish to distinguish them from the finishing ends, so that there may be no mistake when the entire ring is wound in connecting the coils properly to the commutator bars, as cases have happened in which amateurs have, in error, joined both ends of one coil to a segment of the commutator.

As a coil is wound and tested for insulation by a battery and galvanometer, its finishing end should be twisted to the commencement of the next one; it is also important that the recesses between the cogs should contain as nearly as possible the same amount of wire in each, otherwise sparking at the brushes is apt to occur. The cogs, however, on the exterior of the core, and the guides within it (during winding) will equally divide the coils and prevent any strands overlapping. In the employment of wire or plain ring cores without cog divisions on the periphery, the partial winding of one coil over another is apt to burn the coils.

CHAPTER XII.

The operator having satisfied himself that the wire is perfectly insulated from the iron core over which it is wound, the armature can be permanently mounted with the commutator upon the shaft, the nut securing them both tightly in position; where

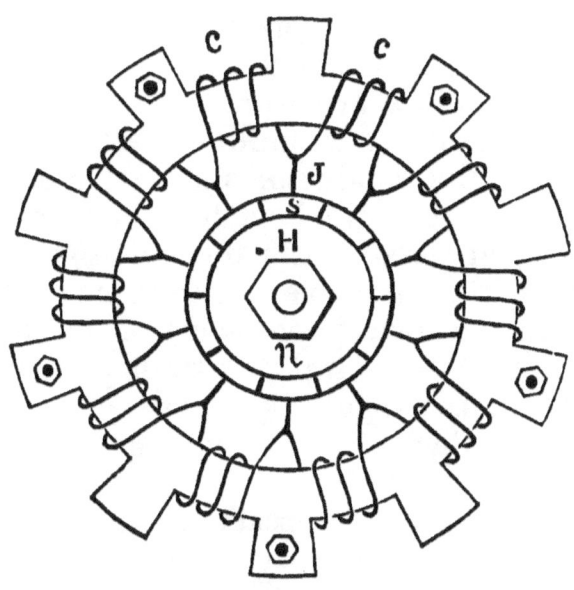

Fig. 23.

the ends of any coils emerge from the ring in close

proximity to the metal arms of the supports, a piece of gutta percha or india rubber tubing should be slipped over the wire for extra insulation.

The ten coils are to be connected in series, and at the point where the finishing end of a coil unites with the beginning of the next one, a junction wire J, Fig. 23, leading from a commutator bar, is also attached. When all joints have been properly effected by soldering, the junctions and wire strands should be painted with Japan black, but if any soldering fluid has been used in making the connections all traces of it should be first removed by applying paraffin oil with a soft camel hair brush, otherwise its corrosive action would soon produce a fault by its destructive effect upon the wire.

In winding the armature core with small wire, great care is required in pulling not to break it; if a severance of the conductor should happen and the fractured ends be still held together by the cotton insulation, increased resistance of the coil would arise, with the probable consequence of greater sparking at the corresponding bar of the commutator when passing the neutral points, or perhaps producing a defective irregularity of the circular form of the segment attached to the damaged coil by a *flat* happening to the bar.

It will now be necessary to bind the armature with two bands of special tinned binding wire, as supplied by the London Electric Wire Company; some tape must first be wrapped upon the outer surface of the coils so that they are not short circuited by the fine binding strands cutting into them. This external wire must be secured by a little solder, which if applied over the cogs does not risk burning the cotton insulation of the wire coils ; in soldering resin only must be used as a flux. An excellent soldering solution for this operation has been patented by Mr. J. H. Watkins, of Eccles. The coils of the armature although tightly wound over the iron if unprotected by the hoop form of binding, would by the action of centrifugal force soon collide with the pole pieces when revolving at high speed, although it is possible that the coils may expand to some extent by a degree of heating, thus contributing to the danger which the external lashing of the armature prevents. The amateur must not however be too lavish in binding the armature coils with a conductor, which is also an insulator of magnetic induction.

WINDING THE FIELD MAGNET.

Having selected the wire for field magnet cores, the next consideration will be the direction of winding : this is shown by the diagram, Fig. 24, and a convenient method is to wind each of the four core bars in separate lengths of wire, afterwards joining the two top commencing or inner ends B B together. The lower commencements B B are also to be united to each other, and the two finishing ends $F.^1 F.^1$ connected together if shunt wound (see Fig. 28). For making these attachments brass double screwed clips as in Fig. 25 will be found very convenient; the two remaining ends F. F. are each connected to a brush, and a wire leading from each brush to binding

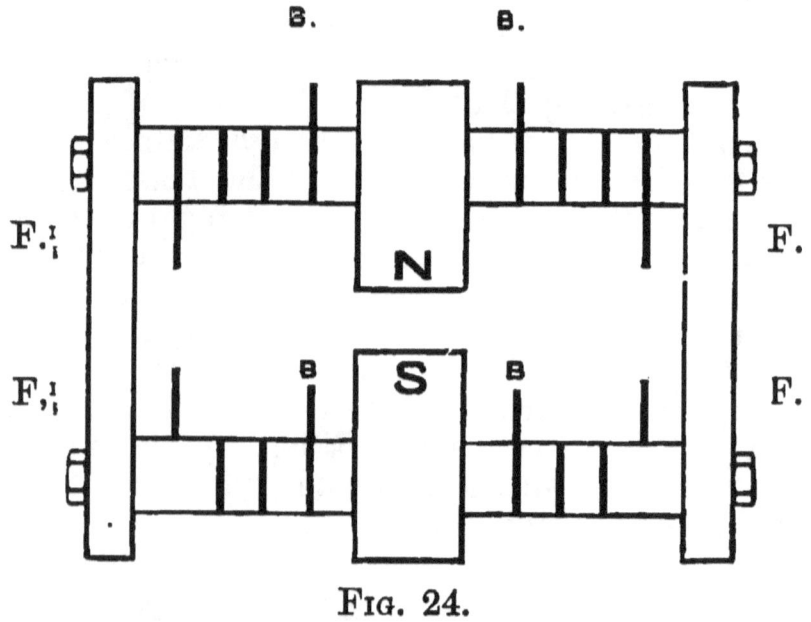

Fig. 24.

screw will complete the arrangement for the shunt system. If, however, the dynamo is series wound,

Fig. 25.

the ends F^1. F^1. must each be joined to a terminal (see Fig. 27) to supply the outer circuit, instead of from the brush clamps as in the shunt wound machine. The amateur possessing a lathe can conveniently wind the cores by mounting them between the centre points and letting the bars revolve slowly by using the back gear if the lathe is so provided. The wire which is fed from a bobbin moving freely on a spindle must be wound tightly and evenly upon the iron cores, and as the last layer upon each limb approaches completion, a few slips of narrow silk ribbon should be wound between the turns for tying and securing the last convolutions from springing undone before the helices have been set fast by some coatings of shellac varnish.

If the dynamo is to be shunt or compound wound, the commencement of each coil should have a short length of much stouter wire soldered to it for greater strength and safety, as there would always be some risk of a small entering wire breaking, and such a mishap would involve unwinding and re-winding to obtain a new attachment to the inner wire; these commencing ends should also have a few inches of small gutta percha or india rubber tubing slipped over them to increase the insulation of the conductors from the iron recesses in the cheeks of pole pieces G, Fig. 2 : suitable tubing can be obtained at any chemist's shop.

When the dynamo is wound, either as shunt or series, it is necessary to impart an initial magnetism to the pole pieces, and any galvanic battery that may be conveniently at hand will do this; the armature may be placed in its position, as its iron toothed ring will act as a keeper to the poles during the process of magnetism, increasing the magnetic intensity; it is, however, advisable not to introduce the resistance of its coils into the magnetising circuit; therefore disconnect the upper and lower field magnet wires which lead to the brush holders and connect them to the binding screws of a detector galvanometer instead.

It is not important that the upper pole piece should be magnetised of North polarity and the bottom one South, as shown in Fig. 23, for it may be *vice versa;* it therefore does not matter which binding screw of the battery is coupled to the right or left terminal of dynamo; the contact of a few seconds will suffice to impart magnetism in the pole pieces by the battery current passing through the four field magnet coils.

The polarity of pole pieces can be ascertained for satisfaction by bringing a poised compass card or galvanometer needle near either pole piece after the battery current has been applied to the circuit; if the upper pole should have acquired north polarity it will attract the south pointing end of the needle; if south, the north pointing end will be deflected towards it.

Now observe the *direction* of deflection of galvanometer needle in the magnetising circuit, and as the immortal Captain Cuttle would remark, *when found make a note of.*

Having denoted deflection of the needle, and supposing the pole pieces to be magnetised N and S as shown in the diagram, Fig. 24, the amateur must understand that when the dynamo is in action, the induced current set up in the armature must produce the same magnetic polarity in these pole pieces as that first imparted to them by the battery process, only in an intensified degree, as the residual magnetism will be but weak after the battery is disconnected from terminals of dynamo, yet sufficient will remain in the iron poles to induce current electricity in the armature when it is revolved between them.

Having successfully magnetised the poles, disconnect the galvanic battery, and it will now be necessary to determine the direction in which the armature must in the future be revolved, and in this experiment its own coils must be brought into action, but not necessarily the field magnet wire.

Do not detach the upper field magnet wire from binding screw of testing galvanometer; the other terminal of this instrument, however, connect to a brush pressing upon the commutator, and the opposite brush join to end of lower field magnet wire, then short circuit the field wires in the manner shewn by a wire spiral at W, Fig. 26, where the connections in this testing circuit comprising the armature A, brushes B, and galvanometer G are represented.

Devise some mechanical arrangement, either from a lathe speed wheel or other source of rotary motion to revolve the armature quickly by a gut cord passing over the pulley on its shaft; a current of electricity will now be induced in the wire coils of the armature through its rapid rotation between the magnetised poles, passing by the commutator bars through brushes thence to the galvanometer, the needle of which will be deflected; and referring to

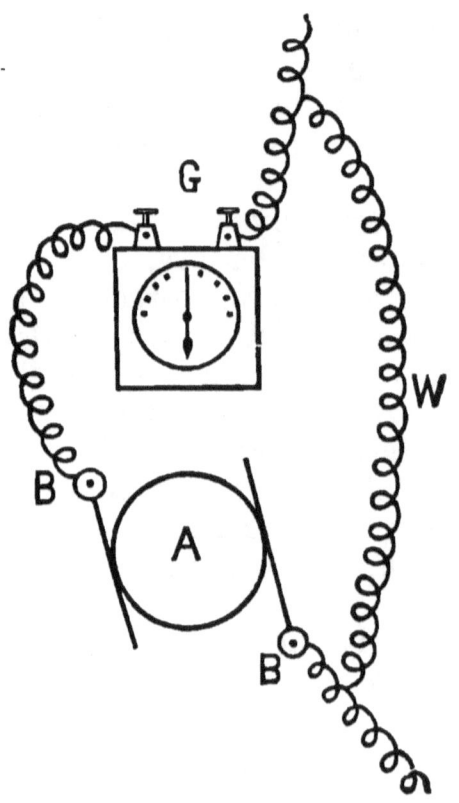

Fig. 26.

the note made upon the historical advice of Captain Cuttle—*if the needle points in the same direction as it did when the battery current was passed through the field coils* the armature has been turned the right way ; if, however, it indicates in the opposite direction, the shaft must be revolved in the reverse manner, so that the revolutions produce the same deflection of galvanometer needle as when a note was made during the process of magnetising by the battery.

The foregoing method has been described in order that the student may clearly understand the principle upon which the armature supplies an electric current to magnetise the pole pieces ; upon starting the dynamo, if it does not immediately light up the lamps it may be presumed that the armature does not furnish current to the electro magnet coils in the right direction, but there is no actual need to reverse the direction of running, as changing over the two field magnet wires, which are connected to the brushes B. B., Fig. 27, will answer the same purpose. And with respect to imparting an inital magnetism to pole-pieces, the Engineer, whose letter appears at page 11, has informed the writer that he immediately started his dynamo without magnetising the poles; this success was undoubtedly due to terrestrial magnetism of the iron, assisted by the carcase being tapped or hammered during the process of fitting together. This instance serves to prove the sensitiveness of the armature described while rotating within a very weak magnetic field produced by the action of the earth upon the iron carcase of the machine.

CHAPTER XIII.

HAVING ascertained the direction in which the armature will turn in the future (which is usually clockwise), remove the galvanometer and let the upper

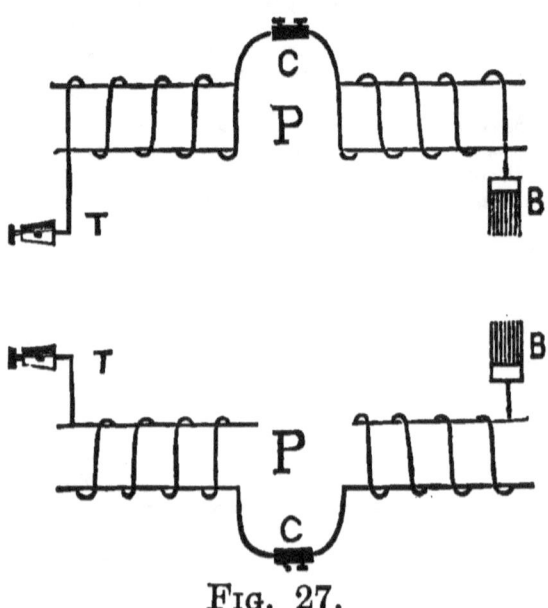

FIG. 27.

and lower field magnet wires be each connected to a spindle supporting a brush; the diagram, Fig. 27, denotes the method of connecting up the wires if series wound; C C clips (see Fig. 25) uniting inner ends of coils bridging over pole pieces P, P. T, T terminals. B, B brushes. Fig. 28 illustrates the circuit if shunt wound.

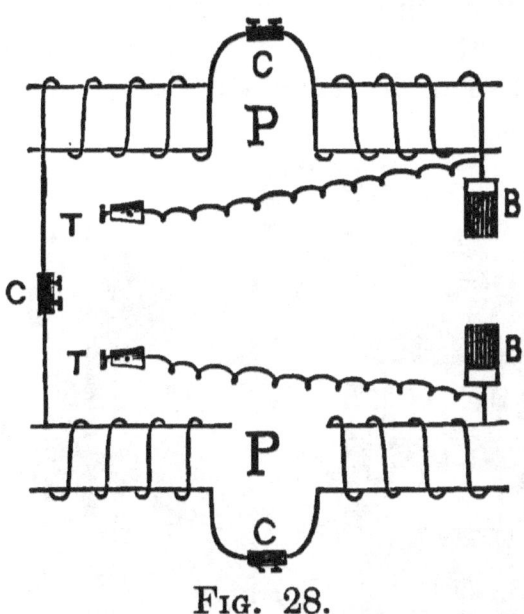

Fig. 28.

It may now be assumed that the dynamo is practically ready for an installation with the two brushes pressing evenly upon the commutator so that they do not jump to cause sparking at the commutator and consequent flickering of the light; the brushes must also be placed at the proper angle of contact with the commutator so that their pressure upon its moving surface is unproductive of noise; they must not be lifted while the dynamo is in action, and must be set to make contact with such bars of the commutator as form the points of current junction of the two halves of the ring diametrically opposite to each other, termed the "neutral points" where the brushes collect the united currents induced in each half of the armature coils.

The neutral points will not be found in a line exactly between the spaces dividing the extremities

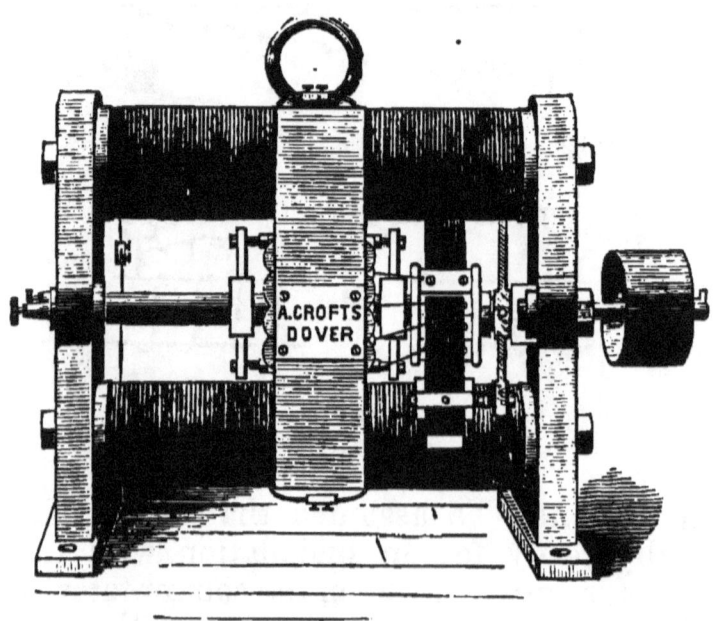

Fig. 29.

of the upper poles from those of the lower, but rather in advance, in the direction of running, and are variable according to the speed of armature rotation and also from any alteration of resistance in lamp circuit, therefore *a lead* has to be given the brushes, which lead requires adjustment upon an increase or decrease of the current generated in armature.

The little Gramme being now completed will weigh about 100 lbs., and the ironwork may be nicely painted with Griffiths' enamel varnish (Oxford blue suits nicely), which dries quickly, and will give it a neat and presentable appearance; the wire coils should receive four or five coats of varnish made up as follows: ¼ pint brown hard varnish, ¼ pint brown polish, 2 oz. Chinese red. Fig. 29 illustrates the finished dynamo, and if the foregoing instructions have been carefully followed out, its efficiency will be about 360 *watts* or 45 volts 8 ampères, a watt being one ampère multiplied by one volt.

A *volt* is the measure for pressure of current, and is approximately equal to the electro-motive force of one Daniell cell.

An *ampère* is the measure for quantity of current.

An *ohm* is the measure of resistance, equal to 129 yards of copper wire $\frac{1}{16}$th of an inch in diameter, but it must be remembered that the resistance of copper wire varies with different makers.

CHAPTER XIV.

MANAGING THE DYNAMO.

The machine should, if possible, be placed in a dry situation, and be kept perfectly clean and free from grease, especially the commutator and brushes; the former may be brightened if dirty, by the application of very fine emery cloth to its surface when slowly revolving, and the brushes can be rinsed in paraffin oil to remove any grease there may be upon them; the ends of both brushes must be evenly cut, and if formed of a bundle of wires should be hollowed concave to suit the circle of the commutator; the copper bands confining the ends must be pushed down as the brushes wear away, but should be kept sufficiently near the ends to prevent any of the wires from straggling outside the commutator—the least touch of oil is sufficient lubrication upon the commutator preparatory to running.

A canvas jacket, oiled and painted, to resist damp, should be made to slip over the dynamo when not in use; to protect it from dust, damp and rust.

DRIVING POWER.

It is important that the driving power should be steady; a half-horse steam, gas, or petroleum engine will drive the dynamo nicely, a rather larger engine, however, will secure a greater steadiness in running; where countershafting is employed, and space permits, the machine should be placed on the opposite side of the motor; the dynamo had better be permanently bolted to an oak slab, and the latter fastened to a floor or solid wood foundation; this method will not risk breaking the projections for the bolts at the base of standards when fixing the machine.

WIRING.

The conducting wires leading from the dynamo to lamps should be sufficiently large so that they may only interpose an insignificant resistance between the generator and the lamps—much smaller wire can, however, be used in attaching each lamp to the mains—the conductors may be of No. 14 wire, covered with vulcanized india rubber, if passing through damp situations; otherwise double cotton covered will answer well enough if dressed with shellac varnish.

The mains should be run in casings of wood, having two grooves or channels planed out an inch or so apart from each other; continuous deal casing can be bought ready prepared for use, at 1½d. per foot, and where practicable this method is preferable to fastening the wire to walls with staples or cleats; an ornamental moulding slip is also sold to screw over the casing containing the conductors, which keeps them snugly out of the way; where small wires bridge the mains for lamps, the junctions should be soldered to the thick wires, so that a faulty joint may not interfere with the brilliancy of a lamp. Strand wires are often used for main conductors covered with vulcanized india rubber, but the cost is considerably more than the single varnished cotton wire.

LAMPS.

The machine if shunt wound will light six of Woodhouse and Rawson's 45 volt 20 candle-power

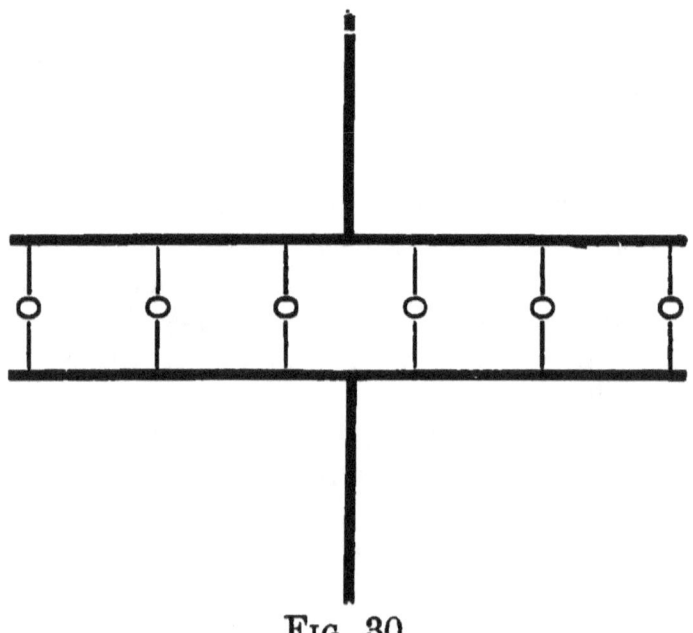

FIG. 30.

lamps, to their full brilliancy if connected in single

parallel system as in diagram, Fig. 30. As the lamps placed nearest to the dynamo will be found brighter than those at the distant end of the circuit, let the leads from terminals of dynamo be connected in the centre of the lamp circuit as shewn, instead of at the extremities of leads, so as to produce a more even distribution of current. The diagram, Fig. 31, illustrates the method of connecting twelve lamps of ten candle-power each in double parallel.

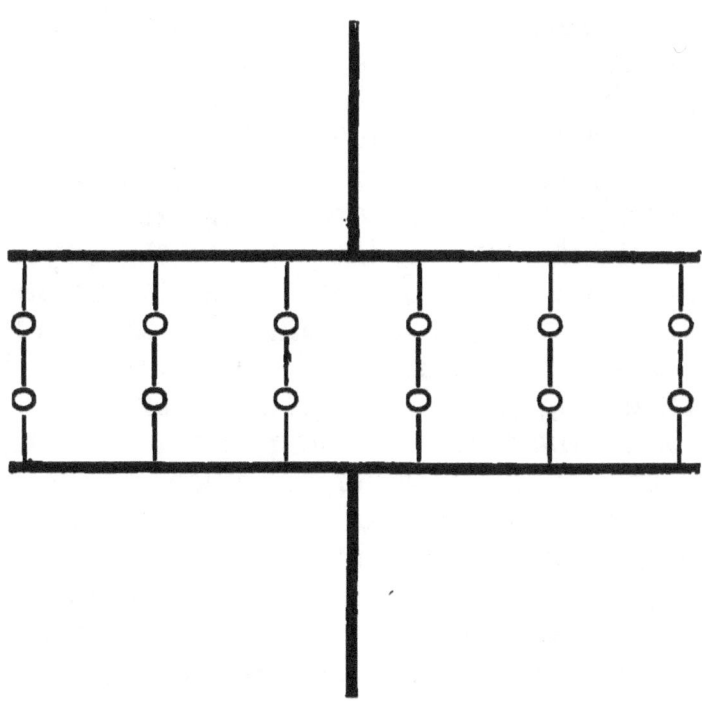

Fig. 31.

CHAPTER XV.

SEMI-INCANDESCENT LAMPS.

In this system of electric lighting, a carbon pencil is arranged to press lightly against a steel or carbon button, causing an intense heating of the carbon point, and small arcs to be formed near the end of the pencil under the button; a splendid light resulting, the carbon pencil is kept up to the button by being supported on an iron float in a column of quicksilver, and the tendency of the float to rise maintains a light and even pressure of the rod against the button; as these lamps are very useful in cases where incandescent or arc lamps are unsuited, and being also within the scope of the amateur to manufacture for himself, an illustration appears in Fig. 32 of this form of lamp, the iron tube below the float not being shown for want of space, the length, however, is given in the references, with other details of construction, so that the reader will have no difficulty in manufacturing them for himself; it is, however, imperative that iron tube should be used, as brass would soon be destroyed by the action of mercury. The lower carbon must be negative, so that the point of the pencil is maintained.

SEMI-INCANDESCENT LAMP.

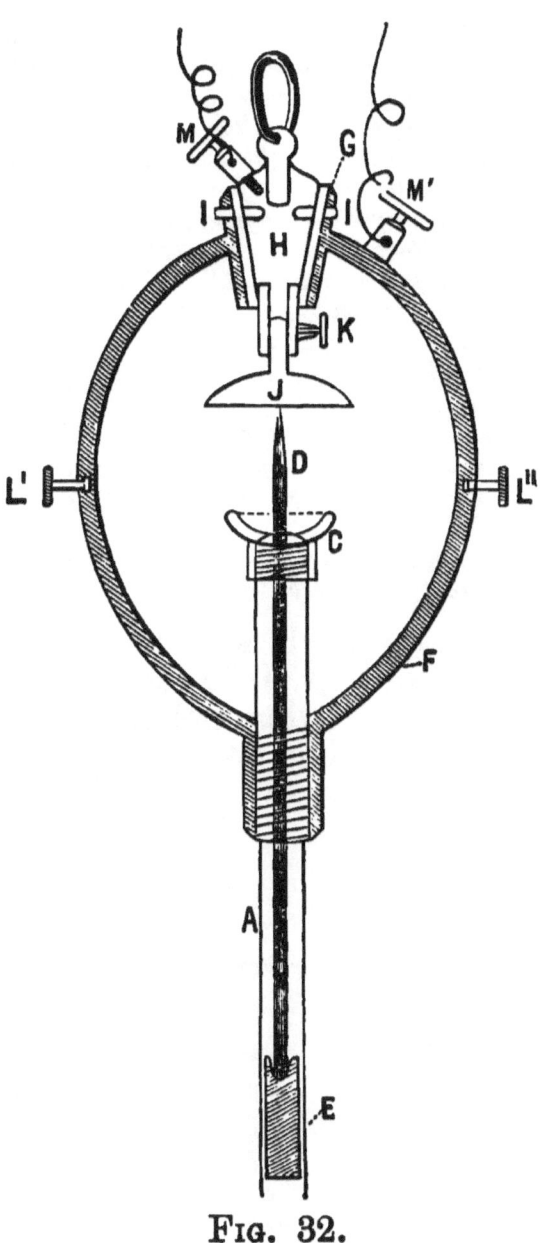

Fig. 32.

References—A, *iron tube $\frac{1}{4}"$ bore, 26" to 30" long; the bore should be smooth as possible, with an iron socket plug screwed on bottom end to close the opening.*

C, *iron cup screwed on top end of tube to contain quicksilver; a hole in the cup admits the carbon pencil passing loosely through it.*

D, *carbon pencil $\frac{1}{8}$ diameter.*

E, *iron float for supporting carbon.*

F, *brass frame jointed at $L'\ L''$, the upper portion supporting the button.*

G, *insulating material.*

H, *brass plug, held in position by the insulating pins 1 o 1.*

J, *carbon or steel button.*

K, *set screw to regulate Carbon button.*

M, *terminal attached to brass plug in electrical contact with button J.*

M', *terminal secured to brass ring frame in contact with mercury and carbon pencil.*

For lighting a single semi-incandescent lamp, small Gramme castings of 60 candle-power can be obtained with the driving pulley grooved to suit a gut cord, for running from a lathe fly-wheel if required.

APPENDIX I.

In the foregoing instructions the dynamo has been regarded as a machine for converting mechanical energy into electricity; a Gramme dynamo, however,

Fig. 33.

will adapt itself for transforming electricity into motive power, the converse of its original purpose;

a small Gramme of 100 candle-power lighting capacity, of which Fig. 33 is an illustration, has been used as a motor to drive sewing machines in a machinist's window near the Cathedral in the City of Canterbury, and was made by the proprietor of the establishment from castings designed by the writer. Mr. Philpott, the maker, considers this little Gramme capable of driving six sewing machines; it has a malleable iron ring casting with sixteen cogs for its armature core and which answers extremely well for the purpose; the following are leading particulars of this little motor :—

Weight	$\frac{3}{4}$ cwt.
Height	13 inches.
Width at base	$8\frac{3}{4}$,,
Extreme length of shaft	$16\frac{1}{4}$,,
Centre of shaft from floor	$6\frac{3}{4}$,,
Width of driving pulley...	2 ,,
Diameter of do.	$3\frac{1}{2}$,,
Do. of armature	$4\frac{7}{8}$,,
Width of do.	$2\frac{1}{4}$,,
Thickness of ring core	$\frac{7}{16}$,,
Inside diameter of do.	$3\frac{1}{4}$,,
Length of each F.M. coil	$3\frac{1}{2}$,,
Diameter of do. (wound) ...	3 ,,
Thickness of F.M. iron core	$1\frac{3}{4}$,,
Width of brushes	$\frac{3}{4}$,,
Weight of armature and shaft	$8\frac{3}{4}$ lbs.
Do. of wound fields with pole pieces	31 ,,
Do. of iron ring core unwound ...	4 ,,
Do. of wire on fields	13 ,,
Do. of ,, armature core ...	$2\frac{1}{2}$,,

It must be remembered that in using the machine

as a motor, the armature will not turn in the same direction as when employed for electric lighting, if it is in series with the field magnet coils. The armature of a shunt motor will, however, revolve the

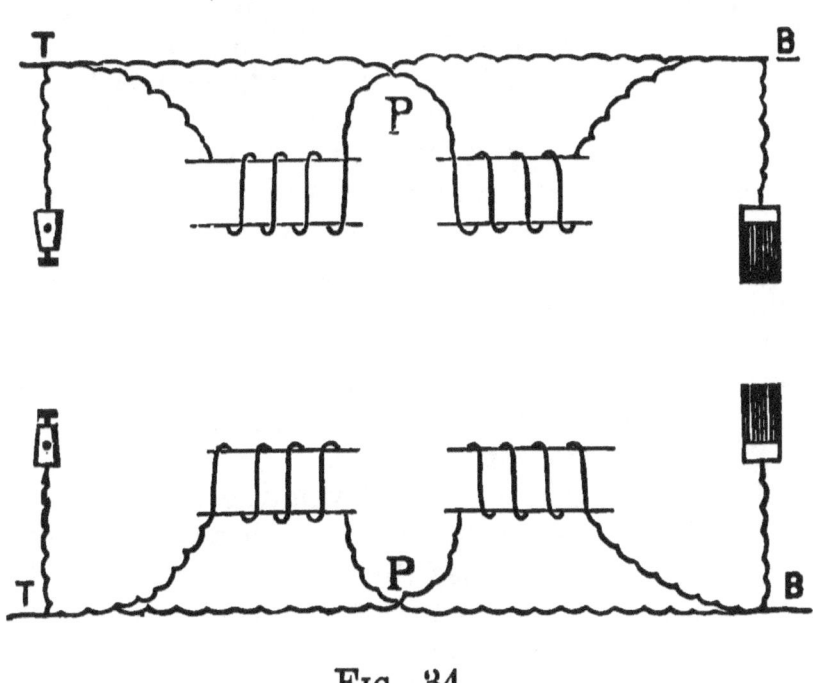

Fig. 34.

References—*P P*, *Wires crossing over pole pieces, but not in contact together.*
T T, Terminal junctions.
B B, Brush junctions.

same way as when the machine is used for generating electricity. As an instance of the performance of a shunt-wound motor, an accomplished electrician,

Mr. Oliver Cornwell, of West Brighton, who is an expert in dynamo engineering, has recently constructed a motor from castings employed in the machine, Fig. 33, the armature being wound with $2\frac{1}{4}$ lbs. of No. 18 double cotton-covered wire, and the field magnet with 10 lbs. of the same size coupled as shunt, and when excited with eight quart bichromate cells it acted most efficiently as a motor, and could not be stopped by hand on the $3\frac{1}{2}''$ pulley; experimenting with it as a series dynamo by coupling the four field coils in multiple system, as shewn in diagram Fig. 34, so as to reduce the resistance of field magnet wire to one quarter the total resistance, it produced a brilliant arc light from $\frac{1}{4}''$ carbons at a speed of 1,800 revolutions, by Mr. Cornwell and two indefatigable assistants treading a heavy lathe-wheel as the source of motive power.

The multiple system of winding and coupling the field magnet coils is often an advantage in a series dynamo, because this method permits of the employment of smaller wire to suit a given resistance of armature; the smaller wire allowing of a greater number of ampère turns on the iron field bars, which magnetises them to saturation, although the current is split or divided through the coils.

By connecting the four coils together *in series* a sufficient resistance is obtained in these small dynamos to admit of their being used as shunt, and by having binding screws on the base board attached to the ends of all field magnet coils with a suitable arrangement of switches, such a motor as Fig. 33, can be quickly converted into either a shunt or series dynamo, and will be found very useful for lighting two 50-candle power semi-incandescent lamps as designed by Mr. Cornwell, and represented in the engraving Fig 32.

The addition of a small fly-wheel on the shaft at opposite end of pulley is a decided advantage to these motors, and will also assist in steadiness of working when used for generating electricity to produce the electric light. It may, however, be remarked that a good dynamo will not always convert itself into a satisfactory motor; a machine that will not act well as a dynamo will often make a splendid motor if its field magnet system is slight and lightly wound.

Engineers and amateurs who are desirous of procuring a complete set of the necessary castings in iron and brass for building the Dynamo herein described, can obtain a Price List of the various parts, also of larger or smaller machines from fifty to five hundred candle power, by enclosing a stamped and addressed envelope to ALFRED CROFTS, *Electrician, Dover; while to those who have finished a dynamo and require some incandescent lamps and fittings, at a moderate price, to complete the installation in an elegant manner, it may be remarked that lamps, &c., can be obtained at Messrs. Wheatley-Kirk-Price and Goulty's Electrical Stores, 39, Queen Street, London, E.C., at a low figure, and it will well repay those who call there, or send for their Catalogue.*

It may also be of interest to aspiring amateurs to read the following extract from a letter dated April 10th, 1890, and received by the writer from an electrician in Oil City, America, to whom a set of 120 c.-p. Gramme castings were sent:—" The machine has been working very satisfactorily. I have seven 40 Volt 16 c.-p. lamps, using a water motor for power."

APPENDIX II.

A Manchester type dynamo carcase is illustrated in Fig. 35, having wrought iron field bars *w w* of

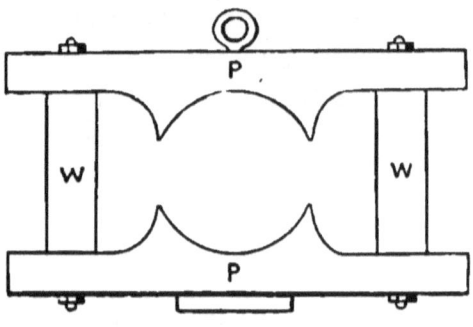

Fig. 35.

round section $2\frac{1}{2}$in. diameter by $5\frac{1}{4}$in. long, uniting two similar castings *p p* by means of nuts screwed upon projecting pins of the forgings *w w* passing through holes in *p p*; the lower casting having extensions at right angles to which plummer blocks are bolted for supporting the armature shaft to a carcase $16\frac{1}{4}$in. long, with a semi-circular space $5\frac{3}{16}$in. by 3in. for receiving a Gramme armature previously described, and containing 8 lbs. of No. 18 wire on each field limb; lighted seven 20 cp lamps brilliantly. As an account of this experiment appeared in the "Dover Chronicle" of February 22nd, 1890, a reprint is appended:

"Several local tradesmen visited Mr. James License's turning factory in St. James' St., on Thursday evening, to examine an installation of the Electric Light that had been fitted up there. A small dynamo and the appliances were made by Mr. CROFTS, a local electrician of some note, and the arrangements were clearly explained by him. A strap was put on to the dynamo machine from one of the pulleys attached to the engine which works Mr. License's lathe and steam saw, and the results were very satisfactory. There were six Edison Swan Lamps, and these fully lit up the workshop. Mr. CROFTS stated that when there was machinery being continually moved by steam or gas engines, the cost of lighting by means of these lamps was very small indeed, whereas the cost of the dynamos of the size shown would not be more than ten or twelve pounds. There is very little doubt Mr. CROFTS has overcome a great difficulty by thus allowing at small cost the electric light to be used on a small scale."

INDEX TO ILLUSTRATIONS.

Professor Faraday .. .. *Frontispiece.*

Fig.   Page

CHAPTER II.

1.—The Gramme Dynamo 17

CHAPTER III.

2.—The Dynamo in Section 22

CHAPTER IV.

3.—Field Magnet Cores and Pole Pieces 25

CHAPTER V.

4.—Side Standard and Brush Support 30
5.—Iron-framed Carcase of Dynamo 32
6.—Gun Metal Bearings for Shaft 34

CHAPTER VI.

7.—Shaft for Armature 37

CHAPTER VII.

8.—Support for Armature	.. ..	40
9.—Iron Stamping for Armature Core	.. .. ..	42
10.— Do. with Bolt Nuts	.. ..	43

CHAPTER VIII.

11.—Commutator Cylinder on Hub..	.. ..	48
12.—Insulated Segments of Commutator	.. .. ..	49

CHAPTER IX.

13.—Adjustable Brush Rocker	.. ..	53
14.— Do. with Brush Clamp attached	.. ..	54
15.—Wire Brush..	.. ..	56
16.—Brush of Flexible Brass Plates	.. .. ..	56
17.—Terminal, for Connecting Main Wires	.. .. ..	57

CHAPTER X.

18.—Diagram of Series Winding	.. ..	62
19.— Do. Shunt „	.. ..	64

CHAPTER XI.

20.—Guides for winding Armature Coils.	.. .. ..	70
21.—Shuttle „ „ „	.. .. ..	71
22.—Armature Core Wound with Wire ..	.. .. ..	73

CHAPTER XII.

23.—Armature Coils Connected to Commutator	.. ..	75
24.—Diagram of Field Magnet Winding..	.. .. ..	78
25.—Method of Coupling Field Magnet Coils..	.. ..	78
26.—How the Armature must Rotate	.. ..	82

CHAPTER XIII.

27.—Coupling Field Wires for Series Winding	84
28.— Do. do. Shunt ,,	85
29.—The Dynamo Completed	86

CHAPTER XIV.

30.—Connecting Lamps in Single Parallel	90
31.— Do. do. Double ,,	91

CHAPTER XV.

32.—Semi-Incandescent Lamp	93

APPENDIX I.

33.—Gramme Motor	95
34.—Multiple Winding of Dynamos	97

APPENDIX II.

35.—The Manchester Dynamo	100

7, STATIONERS' HALL COURT, LONDON, E.C.
January, 1888.

A CATALOGUE OF BOOKS

INCLUDING MANY NEW AND STANDARD WORKS IN

ENGINEERING, MECHANICS, ARCHITECTURE,

NATURAL AND APPLIED SCIENCE,

INDUSTRIAL ARTS, TRADE AND COMMERCE, AGRICULTURE,

GARDENING, LAND MANAGEMENT, LAW, &c.

PUBLISHED BY

CROSBY LOCKWOOD & SON.

MECHANICS, MECHANICAL ENGINEERING, etc.

New Manual for Practical Engineers.

THE PRACTICAL ENGINEER'S HAND-BOOK. Comprising a Treatise on Modern Engines and Boilers, Marine, Locomotive and Stationary. And containing a large collection of Rules and Practical Data relating to recent Practice in Designing and Constructing all kinds of Engines, Boilers, and other Engineering work. The whole constituting a comprehensive Key to the Board of Trade and other Examinations for Certificates of Competency in Modern Mechanical Engineering. By WALTER S. HUTTON, Civil and Mechanical Engineer, Author of "The Works' Manager's Hand-book for Engineers," &c. With upwards of 370 Illustrations. Second Edition, Revised with Additions. Medium 8vo, nearly 500 pp., price 18s. Strongly bound. [*Just published.*

☞ *This work is designed as a companion to the Author's* "WORKS' MANAGER'S HAND-BOOK." *It possesses many new and original features, and contains, like its predecessor, a quantity of matter not originally intended for publication, but collected by the author for his own use in the construction of a great variety of modern engineering work.*

The information is given in a condensed and concise form, and is illustrated by upwards of 370 Woodcuts; and comprises a quantity of tabulated matter of great value to all engaged in designing, constructing, or estimating for ENGINES, BOILERS *and* OTHER ENGINEERING WORK.

*** OPINIONS OF THE PRESS.

"We have kept it at hand for several weeks, referring to it as occasion arose, and we have not on a single occasion consulted its pages without finding the information of which we were in quest."—*Athenæum*.

"A thoroughly good practical handbook, which no engineer can go through without learning something that will be of service to him."—*Marine Engineer*.

"An excellent book of reference for engineers, and a valuable text-book for students of engineering."—*Scotsman*.

"This valuable manual embodies the results and experience of the leading authorities on mechanical engineering."—*Building News*.

"The author has collected together a surprising quantity of rules and practical data, and has shown much judgment in the selections he has made. . . . There is no doubt that this book is one of the most useful of its kind published, and will be a very popular compendium."—*Engineer*.

"A mass of information, set down in simple language, and in such a form that it can be easily referred to at any time. The matter is uniformly good and well chosen, and is greatly elucidated by the illustrations. The book will find its way on to most engineers' shelves, where it will rank as one of the most useful books of reference."—*Practical Engineer*.

"Full of useful information, and should be found on the office shelf of all practical engineers."—*English Mechanic*.

Handbook for Works' Managers.

THE WORKS' MANAGER'S HANDBOOK OF MODERN RULES, TABLES, AND DATA. For Engineers, Millwrights, and Boiler Makers; Tool Makers, Machinists, and Metal Workers; Iron and Brass Founders, &c. By W. S. HUTTON, Civil and Mechanical Engineer, Author of "The Practical Engineer's Handbook." Third Edition, carefully Revised, with Additions. In One handsome Vol., medium 8vo, price 15s. strongly bound.

☞ *The Author having compiled Rules and Data for his own use in a great variety of modern engineering work, and having found his notes extremely useful, decided to publish them—revised to date—believing that a practical work, suited to the* DAILY REQUIREMENTS OF MODERN ENGINEERS, *would be favourably received.*

In the Third Edition, the following among other additions have been made, viz.: Rules for the Proportions of Riveted Joints in Soft Steel Plates, the Results of Experiments by PROFESSOR KENNEDY *for the Institution of Mechanical Engineers—Rules for the Proportions of Turbines—Rules for the Strength of Hollow Shafts of Whitworth's Compressed Steel, &c.*

*** OPINIONS OF THE PRESS.

"The author treats every subject from the point of view of one who has collected workshop notes for application in workshop practice, rather than from the theoretical or literary aspect. The volume contains a great deal of that kind of information which is gained only by practical experience, and is seldom written in books."—*Engineer.*

"The volume is an exceedingly useful one, brimful with engineers' notes, memoranda, and rules, and well worthy of being on every mechanical engineer's bookshelf."—*Mechanical World.*

"A formidable mass of facts and figures, readily accessible through an elaborate index . . . . Such a volume will be found absolutely necessary as a book of reference in all sorts of 'works' connected with the metal trades."—*Ryland's Iron Trades Circular.*

"Brimful of useful information, stated in a concise form, Mr. Hutton's books have met a pressing want among engineers. The book must prove extremely useful to every practical man possessing a copy."—*Practical Engineer.*

The Modernised "Templeton."

THE PRACTICAL MECHANIC'S WORKSHOP COMPANION. Comprising a great variety of the most useful Rules and Formulæ in Mechanical Science, with numerous Tables of Practical Data and Calculated Results for Facilitating Mechanical Operations. By WILLIAM TEMPLETON, Author of "The Engineer's Practical Assistant," &c. &c. Fifteenth Edition, Revised, Modernised, and considerably Enlarged by WALTER S. HUTTON, C.E., Author of "The Works' Manager's Handbook," "The Practical Engineer's Handbook," &c. Fcap. 8vo, nearly 500 pp., with Eight Plates and upwards of 250 Illustrative Diagrams, 6s., strongly bound for workshop or pocket wear and tear.

☞ TEMPLETON'S "MECHANIC'S WORKSHOP COMPANION" *has been for more than a quarter of a century deservedly popular, and, as a recognised Text-Book and well-worn and thumb-marked* vade mecum *of several generations of intelligent and aspiring workmen, it has had the reputation of having been the means of raising many of them in their position in life.*

In consequence of the lapse of time since the Author's death, and the great advances in Mechanical Science, the Publishers have thought it advisable to have it entirely Reconstructed and Modernised; and in its present greatly Enlarged and Improved form, they are sure that it will commend itself to the English workmen of the present day all the world over, and become, like its predecessors, their indispensable friend and referee.

A smaller type having been adopted, and the page increased in size, while the number of pages has advanced from about 330 to nearly 500, the book practically contains double the amount of matter that was comprised in the original work.

*** OPINIONS OF THE PRESS.

"In its modernised form Hutton's 'Templeton' should have a wide sale, for it contains much valuable information which the mechanic will often find of use, and not a few tables and notes which he might look for in vain in other works. This modernised edition will be appreciated by all who have learned to value the original editions of 'Templeton.'"—*English Mechanic.*

"It has met with great success in the engineering workshop, as we can testify; and there are a great many men who, in a great measure, owe their rise in life to this little book."—*Building News.*

"This familiar text-book—well known to all mechanics and engineers—is of essential service to the every-day requirements of engineers, millwrights and the various trades connected with engineering and building. The new modernised edition is worth its weight in gold."—*Building News.* (Second Notice.)

"The publishers wisely entrusted the task of revision of this popular, valuable and useful book to Mr. Hutton, than whom a more competent man they could not have found."—*Iron.*

Stone-working Machinery.

STONE-WORKING MACHINERY, and the Rapid and Economical Conversion of Stone. With Hints on the Arrangement and Management of Stone Works. By M. POWIS BALE, M.I.M.E., A.M.I.C.E. With numerous Illustrations. Large crown 8vo, 9s. cloth.

"The book should be in the hands of every mason or student of stone-work."—*Colliery Guardian.*

"It is in every sense of the word a standard work upon a subject which the author is fully competent to deal exhaustively with."—*Builder's Weekly Reporter.*

"A capital handbook for all who manipulate stone for building or ornamental purposes."—*Machinery Market.*

Turning.

LATHE-WORK : A Practical Treatise on the Tools, Appliances, and Processes employed in the Art of Turning. By PAUL N. HASLUCK. Third Edition, Revised and Enlarged. Crown 8vo, 5s. cloth.

"Written by a man who knows, not only how work ought to be done, but who also knows how to do it, and how to convey his knowledge to others. To all turners this book would be valuable."—*Engineering.*

"We can safely recommend the work to young engineers. To the amateur it will simply be invaluable. To the student it will convey a great deal of useful information."—*Engineer.*

"A compact, succinct, and handy guide to lathe-work did not exist in our language until Mr. Hasluck, by the publication of this treatise, gave the turner a true *vade-mecum.*"—*House Decorator*

Screw-Cutting.

SCREW THREADS : And Methods of Producing Them. With Numerous Tables, and complete directions for using Screw-Cutting Lathes. By PAUL N. HASLUCK, Author of "Lathe-Work," "The Metal Turner's Handybook," &c. Waistcoat-pocket size, price 1s. [*Just published.*

"Full of useful information, hints and practical criticism. Taps, dies and screwing-tools generally are illustrated and their action described."—*Mechanical World.*

"It is a complete compendium of all the details of the screw-cutting lathe; in fact, a *multum in parvo* on all the subjects it treats upon."—*Carpenter and Builder.*

Engineer's and Machinist's Assistant.

THE ENGINEER'S, MILLWRIGHT'S, and MACHINIST'S PRACTICAL ASSISTANT. A collection of Useful Tables, Rules and Data. By WILLIAM TEMPLETON. Seventh Edition, with Additions. 18mo, 2s. 6d. cloth.

"Templeton's handbook occupies a foremost place among books of this kind. A more suitable present to an apprentice to any of the mechanical trades could not possibly be made."—*Building News.*

"A deservedly appreciated work, which should be in the 'drawer' of every mechanic."—*English Mechanic.*

Iron and Steel.

"IRON AND STEEL" : A Work for the Forge, Foundry, Factory, and Office. Containing ready, useful, and trustworthy Information for Ironmasters and their Stock-takers; Managers of Bar, Rail, Plate, and Sheet Rolling Mills; Iron and Metal Founders; Iron Ship and Bridge Builders; Mechanical, Mining, and Consulting Engineers; Architects, Contractors, Builders, and Professional Draughtsmen. By CHARLES HOARE, Author of "The Slide Rule," &c. Eighth Edition, Revised throughout and considerably Enlarged. With folding Scales of "Foreign Measures compared with the English Foot," and "Fixed Scales of Squares, Cubes, and Roots, Areas Decimal Equivalents," &c. Oblong 32mo, leather, elastic band, 6s.

"For comprehensiveness the book has not its equal."—*Iron.*

"One of the best of the pocket books, and a useful companion in other branches of work than iron and steel."—*English Mechanic.*

"We cordially recommend this book to those engaged in considering the details of all kinds of iron and steel works."—*Naval Science.*

High-Pressure Steam Engines.

THE HIGH-PRESSURE STEAM-ENGINE : An Exposition of its Comparative Merits and an Essay towards an Improved System of Construction. By Dr. ERNST ALBAN. Translated from the German, with Notes, by Dr. POLE, M. Inst. C.E., &c. With 28 Plates. 8vo, 16s. 6d. cloth.

"Goes thoroughly into the examination of the high-pressure engine, the boiler, and its appendages and deserves a place in every scientific library."—*Steam Shipping Chronicle.*

Engineering Construction.

PATTERN-MAKING: *A Practical Treatise*, embracing the Main Types of Engineering Construction, and including Gearing, both Hand and Machine made, Engine Work, Sheaves and Pulleys, Pipes and Columns, Screws, Machine Parts, Pumps and Cocks, the Moulding of Patterns in Loam and Greensand, &c., together with the methods of Estimating the weight of Castings; to which is added an Appendix of Tables for Workshop Reference. By a FOREMAN PATTERN MAKER. With upwards of Three Hundred and Seventy Illustrations. Crown 8vo, 7s. 6d. cloth.

"A well-written technical guide, evidently written by a man who understands and has practised what he has written about. We cordially recommend it to engineering students, young journeymen, and others desirous of being initiated into the mysteries of pattern-making."—*Builder.*

"Likely to prove a welcome guide to many workmen, especially to draughtsmen who have lacked a training in the shops, pupils pursuing their practical studies in our factories, and to employers and managers in engineering works."—*Hardware Trade Journal.*

"More than 370 illustrations help to explain the text, which is, however, always clear and explicit, thus rendering the work an excellent *vade mecum* for the apprentice who desires to become master of his trade."—*English Mechanic.*

Dictionary of Mechanical Engineering Terms.

LOCKWOOD'S DICTIONARY OF TERMS USED IN THE PRACTICE OF MECHANICAL ENGINEERING, embracing those current in the Drawing Office, Pattern Shop, Foundry, Fitting, Turning, Smith's and Boiler Shops, &c. &c. Comprising upwards of 6,000 Definitions. Edited by A FOREMAN PATTERN-MAKER, Author of "Pattern Making." Crown 8vo, 7s. 6d. cloth. [*Just published.*

"Just the sort of handy dictionary required by the various trades engaged in mechanical engineering. The practical engineering pupil will find the book of great value in his studies, and every foreman engineer and mechanic should have a copy."—*Building News.*

"After a careful examination of the book, and trying all manner of words, we think that the engineer will here find all he is likely to require. It will be largely used."—*Practical Engineer.*

"This admirable dictionary, although primarily intended for the use of draughtsmen and other technical craftsmen, is of much larger value as a book of reference, and will find a ready welcome in many libraries."—*Glasgow Herald.*

Smith's Tables for Mechanics, etc.

TABLES, MEMORANDA, AND CALCULATED RESULTS, FOR MECHANICS, ENGINEERS, ARCHITECTS, BUILDERS, etc. Selected and Arranged by FRANCIS SMITH. Fourth Edition, Revised and Enlarged, 250 pp., waistcoat-pocket size, 1s. 6d. limp leather. [*Just published.*

"It would, perhaps, be as difficult to make a small pocket-book selection of notes and formulæ to suit ALL engineers as it would be to make a universal medicine; but Mr. Smith's waistcoat-pocket collection may be looked upon as a successful attempt."—*Engineer.*

"The best example we have ever seen of 250 pages of useful matter packed into the dimensions of a card-case."—*Building News.*

"A veritable pocket treasury of knowledge.'—*Iron.*

Steam Boilers.

A TREATISE ON STEAM BOILERS: Their Strength, Construction, and Economical Working. By ROBERT WILSON, C.E. Fifth Edition. 12mo, 6s. cloth.

"The best treatise that has ever been published on steam boilers."—*Engineer.*

"The author shows himself perfect master of his subject, and we heartily recommend all employing steam power to possess themselves of the work."—*Ryland's Iron Trade Circular.*

Boiler Chimneys.

BOILER AND FACTORY CHIMNEYS: Their Draught-Power and Stability. With a Chapter on *Lightning Conductors.* By ROBERT WILSON, A.I.C.E., Author of "A Treatise on Steam Boilers," &c. Second Edition. Crown 8vo, 3s. 6d. cloth. [*Just published.*

Boiler Making.

THE BOILER-MAKER'S READY RECKONER. With Examples of Practical Geometry and Templating, for the Use of Platers, Smiths and Riveters. By JOHN COURTNEY, Edited by D. K. CLARK, M.I.C.E. Second Edition, Revised, with Additions, 12mo, 5s. half-bound.

"A most useful work. . . . . No workman or apprentice should be without this book."—*Iron Trade Circular.*

"A reliable guide to the working boiler-maker."—*Iron.*

"Boiler-makers will readily recognise the value of this volume. . . . The tables are clearly printed, and so arranged that they can be referred to with the greatest facility, so that it cannot be doubted that they will be generally appreciated and much used."—*Mining Journal.*

Steam Engine.

TEXT-BOOK ON THE STEAM ENGINE. With a Supplement on Gas Engines. By T. M. GOODEVE, M.A., Barrister-at-Law, Author of "The Elements of Mechanism," &c. Tenth Edition. With numerous Illustrations. Crown 8vo, 6s. cloth.

"Professor Goodeve has given us a treatise on the steam engine which will bear comparison with anything written by Huxley or Maxwell, and we can award it no higher praise."—*Engineer.*

"Professor Goodeve's book is ably and clearly written. It is a sound work."—*Athenæum.*

"Mr. Goodeve's text-book is a work of which every young engineer should possess himself." —*Mining Journal.*

"Essentially practical in its aims. The manner of exposition leaves nothing to be desired."— *Scotsman.*

"A valuable *vade mecum* for the student of engineering and should be in the possession of every scientific reader."—*Colliery Guardian.*

Gas Engines.

ON GAS-ENGINES. Being a Reprint, with some Additions, of the Supplement to the *Text-book on the Steam Engine*, by T. M. GOODEVE, M.A. Crown 8vo, 2s. 6d. cloth.

"Like all Mr. Goodeve's writings, the present is no exception in point of general excellence It is a valuable little volume."—*Mechanical World.*

"This little book will be useful to those who desire to understand how the gas-engine works." —*English Mechanic.*

Steam.

THE SAFE USE OF STEAM. Containing Rules for Unprofessional Steam-users. By an ENGINEER. Sixth Edition. Sewed, 6d.

"If steam-users would but learn this little book by heart boiler explosions would become sensations by their rarity."—*English Mechanic.*

Coal and Speed Tables.

A POCKET BOOK OF COAL AND SPEED TABLES, for Engineers and Steam-users. By NELSON FOLEY, Author of "Boiler Construction." Pocket-size, 3s. 6d. cloth; 4s. leather.

"This is a very useful book, containing very useful tables. The results given are well chosen, and the volume contains evidence that the author really understands his subject. We can recommend the work with pleasure."—*Mechanical World.*

"These tables are designed to meet the requirements of every-day use; they are of sufficient scope for most practical purposes, and may be commended to engineers and users of steam."— *Iron.*

"This pocket-book well merits the attention of the practical engineer. Mr. Foley has compiled a very useful set of tables, the information contained in which is frequently required by engineers, coal consumers and users of steam."—*Iron and Coal Trades Review.*

Fire Engineering.

FIRES, FIRE-ENGINES, AND FIRE-BRIGADES. With a History of Fire-Engines, their Construction, Use, and Management; Remarks on Fire-Proof Buildings, and the Preservation of Life from Fire; Statistics of the Fire Appliances in English Towns; Foreign Fire Systems; Hints on Fire Brigades, &c. &c. By CHARLES F. T. YOUNG, C.E. With numerous Illustrations, 544 pp., demy 8vo, £1 4s. cloth.

"To such of our readers as are interested in the subject of fires and fire apparatus, we can most heartily commend this book. It is really the only English work we now have upon the subject."— *Engineering.*

"It displays much evidence of careful research; and Mr. Young has put his facts neatly together. It is evident enough that his acquaintance with the practical details of the construction of steam fire engines, old and new, and the conditions with which it is necessary they should comply, is accurate and full."—*Engineer.*

Gas Lighting.

COMMON SENSE FOR GAS-USERS: A Catechism of Gas-Lighting for *Householders, Gasfitters, Millowners, Architects, Engineers, etc.* By ROBERT WILSON, C.E., Author of "A Treatise on Steam Boilers." Second Edition, with Folding Plates and Wood Engravings. Crown 8vo, price 1s. in wrapper.

"All gas-users will decidedly benefit, both in pocket and comfort, if they will avail themselves of Mr. Wilson's counsels."—*Engineering.*

THE POPULAR WORKS OF MICHAEL REYNOLDS
("The Engine Driver's Friend").

Locomotive-Engine Driving.
LOCOMOTIVE-ENGINE DRIVING: A Practical Manual for Engineers in charge of Locomotive Engines. By Michael Reynolds, Member of the Society of Engineers, formerly Locomotive Inspector L. B. and S. C. R. Eighth Edition. Including a Key to the Locomotive Engine. With Illustrations and Portrait of Author. Crown 8vo, 4s. 6d. cloth.

"Mr. Reynolds has supplied a want, and has supplied it well. We can confidently recommend the book, not only to the practical driver, but to everyone who takes an interest in the performance of locomotive engines."—*The Engineer.*

"Mr. Reynolds has opened a new chapter in the literature of the day. This admirable practical treatise, of the practical utility of which we have to speak in terms of warm commendation."—*Athenæum.*

"Evidently the work of one who knows his subject thoroughly."—*Railway Service Gazette.*

"Were the cautions and rules given in the book to become part of the every-day working of our engine-drivers, we might have fewer distressing accidents to deplore."—*Scotsman.*

Stationary Engine Driving.
STATIONARY ENGINE DRIVING: A Practical Manual for Engineers in charge of Stationary Engines. By Michael Reynolds. Third Edition, Enlarged. With Plates and Woodcuts. Crown 8vo, 4s. 6d. cloth.

"The author is thoroughly acquainted with his subjects, and his advice on the various points treated is clear and practical. . . . He has produced a manual which is an exceedingly useful one for the class for whom it is specially intended."—*Engineering.*

"Our author leaves no stone unturned. He is determined that his readers shall not only know something about the stationary engine, but all about it."—*Engineer.*

"An engineman who has mastered the contents of Mr. Reynolds's book will require but little actual experience with boilers and engines before he can be trusted to look after them."—*English Mechanic*

The Engineer, Fireman, and Engine-Boy.
THE MODEL LOCOMOTIVE ENGINEER, FIREMAN, and ENGINE-BOY. Comprising a Historical Notice of the Pioneer Locomotive Engines and their Inventors. By Michael Reynolds. With numerous Illustrations and a fine Portrait of George Stephenson. Crown 8vo, 4s. 6d. cloth.

"From the technical knowledge of the author it will appeal to the railway man of to-day more forcibly than anything written by Dr. Smiles. . . . The volume contains information of a technical kind, and facts that every driver should be familiar with."—*English Mechanic.*

"We should be glad to see this book in the possession of everyone in the kingdom who has ever laid, or is to lay, hands on a locomotive engine."—*Iron.*

Continuous Railway Brakes.
CONTINUOUS RAILWAY BRAKES: A Practical Treatise on the several Systems in Use in the United Kingdom; their Construction and Performance. With copious Illustrations and numerous Tables. By Michael Reynolds. Large crown 8vo, 9s. cloth.

"A popular explanation of the different brakes. It will be of great assistance in forming public opinion, and will be studied with benefit by those who take an interest in the brake."—*English Mechanic.*

"Written with sufficient technical detail to enable the principle and relative connection of the various parts of each particular brake to be readily grasped."—*Mechanical World.*

Engine-Driving Life.
ENGINE-DRIVING LIFE; or, Stirring Adventures and Incidents in the Lives of Locomotive-Engine Drivers. By Michael Reynolds. Ninth Thousand. Crown 8vo, 2s. cloth.

"From first to last is perfectly fascinating. Wilkie Collins's most thrilling conceptions are thrown into the shade by true incidents, endless in their variety, related in every page."—*North British Mail.*

"Anyone who wishes to get a real insight into railway life cannot do better than read 'Engine-Driving Life' for himself; and if he once take it up he will find that the author's enthusiasm and real love of the engine-driving profession will carry him on till he has read every page."—*Saturday Review.*

Pocket Companion for Enginemen.
THE ENGINEMAN'S POCKET COMPANION AND PRACTICAL EDUCATOR FOR ENGINEMEN, BOILER ATTENDANTS, AND MECHANICS. By Michael Reynolds. With Forty-five Illustrations and numerous Diagrams. Second Edition, Revised. Royal 18mo, 3s. 6d., strongly bound for pocket wear. [*Just published.*

"This admirable work is well suited to accomplish its object, being the honest workmanship of a competent engineer."—*Glasgow Herald.*

"A most meritorious work, giving in a succinct and practical form all the information an engineminder desirous of mastering the scientific principles of his daily calling would require."—*Miller.*

"A boon to those who are striving to become efficient mechanics."—*Daily Chronicle.*

French-English Glossary for Engineers, etc.

A POCKET GLOSSARY of TECHNICAL TERMS: ENGLISH-FRENCH, FRENCH-ENGLISH; with Tables suitable for the Architectural, Engineering, Manufacturing and Nautical Professions. By JOHN JAMES FLETCHER, Engineer and Surveyor; 200 pp. Waistcoat-pocket size, 1s. 6d., limp leather.

"It ought certainly to be in the waistcoat-pocket of every professional man."—*Iron.*
"It is a very great advantage for readers and correspondents in France and England to have so large a number of the words relating to engineering and manufacturers collected in a lilliputian volume. The little book will be useful both to students and travellers."—*Architect.*
"The glossary of terms is very complete, and many of the tables are new and well arranged. We cordially commend the book."—*Mechanical World.*

Portable Engines.

THE PORTABLE ENGINE; ITS CONSTRUCTION AND MANAGEMENT. A Practical Manual for Owners and Users of Steam Engines generally. By WILLIAM DYSON WANSBROUGH. With 90 Illustrations. Crown 8vo, 3s. 6d. cloth. [*Just published.*

"This is a work of value to those who use steam machinery. . . . Should be read by everyone who has a steam engine, on a farm or elsewhere."—*Mark Lane Express.*
"We cordially commend this work to buyers and owners of steam engines, and to those who have to do with their construction or use."—*Timber Trades Journal.*
"Such a general knowledge of the steam engine as Mr. Wansbrough furnishes to the reader should be acquired by all intelligent owners and others who use the steam engine."—*Building News.*

CIVIL ENGINEERING, SURVEYING, etc.

MR. HUMBER'S IMPORTANT ENGINEERING BOOKS.

The Water Supply of Cities and Towns.

A COMPREHENSIVE TREATISE on the WATER-SUPPLY OF CITIES AND TOWNS. By WILLIAM HUMBER, A-M. Inst. C.E., and M. Inst. M.E., Author of "Cast and Wrought Iron Bridge Construction," &c. &c. Illustrated with 50 Double Plates, 1 Single Plate, Coloured Frontispiece, and upwards of 250 Woodcuts, and containing 400 pages of Text. Imp. 4to, £6 6s. elegantly and substantially half-bound in morocco.

List of Contents.

I. Historical Sketch of some of the means that have been adopted for the Supply of Water to Cities and Towns.—II. Water and the Foreign Matter usually associated with it.—III. Rainfall and Evaporation.—IV. Springs and the water-bearing formations of various districts.—V. Measurement and Estimation of the flow of Water.—VI. On the Selection of the Source of Supply.—VII. Wells.—VIII. Reservoirs.—IX. The Purification of Water.—X. Pumps. — XI. Pumping Machinery. — XII. Conduits.—XIII. Distribution of Water.—XIV. Meters, Service Pipes, and House Fittings.—XV. The Law and Economy of Water Works.—XVI. Constant and Intermittent Supply.—XVII. Description of Plates. — Appendices, giving Tables of Rates of Supply, Velocities, &c. &c., together with Specifications of several Works illustrated, among which will be found: Aberdeen, Bideford, Canterbury, Dundee, Halifax, Lambeth, Rotherham, Dublin, and others.

"The most systematic and valuable work upon water supply hitherto produced in English, or in any other language. . . . Mr. Humber's work is characterised almost throughout by an exhaustiveness much more distinctive of French and German than of English technical treatises."—*Engineer.*
"We can congratulate Mr. Humber on having been able to give so large an amount of information on a subject so important as the water supply of cities and towns. The plates, fifty in number, are mostly drawings of executed works, and alone would have commanded the attention of every engineer whose practice may lie in this branch of the profession."—*Builder.*

Cast and Wrought Iron Bridge Construction.

A COMPLETE AND PRACTICAL TREATISE ON CAST AND WROUGHT IRON BRIDGE CONSTRUCTION, including Iron Foundations. In Three Parts—Theoretical, Practical, and Descriptive. By WILLIAM HUMBER, A-M. Inst. C.E., and M. Inst. M.E. Third Edition, Revised and much improved, with 115 Double Plates (20 of which now first appear in this edition), and numerous Additions to the Text. In Two Vols., imp. 4to, £6 16s. 6d. half-bound in morocco.

"A very valuable contribution to the standard literature of civil engineering. In addition to elevations, plans and sections, large scale details are given which very much enhance the instructive worth of these illustrations."—*Civil Engineer and Architect's Journal.*
"Mr. Humber's stately volumes, lately issued—in which the most important bridges erected during the last five years, under the direction of the late Mr. Brunel, Sir W. Cubitt, Mr. Hawkshaw, Mr. Page, Mr. Fowler, Mr. Hemans, and others among our most eminent engineers, are drawn and specified in great detail."—*Engineer.*

MR. HUMBER'S GREAT WORK ON MODERN ENGINEERING.

Complete in Four Volumes, imperial 4to, price £12 12s., half-morocco. Each Volume sold separately as follows:—

A RECORD OF THE PROGRESS OF MODERN ENGINEERING. FIRST SERIES.
Comprising Civil, Mechanical, Marine, Hydraulic, Railway, Bridge, and other Engineering Works, &c. By WILLIAM HUMBER, A-M. Inst. C.E., &c. Imp. 4to, with 36 Double Plates, drawn to a large scale, Photographic Portrait of John Hawkshaw, C.E., F.R.S., &c., and copious descriptive Letterpress, Specifications, &c., £3 3s. half-morocco.

List of the Plates and Diagrams.

Victoria Station and Roof, L. B. & S. C. R. (8 plates); Southport Pier (2 plates); Victoria Station and Roof, L. C. & D. and G. W. R. (6 plates); Roof of Cremorne Music Hall; Bridge over G. N. Railway; Roof of Station, Dutch Rhenish Rail (2 plates); Bridge over the Thames, West London Extension Railway (5 plates); Armour Plates: Suspension Bridge, Thames (4 plates); The Allen Engine; Suspension Bridge, Avon (3 plates); Underground Railway (3 plates).

"Handsomely lithographed and printed. It will find favour with many who desire to preserve in a permanent form copies of the plans and specifications prepared for the guidance of the contractors for many important engineering works."—*Engineer.*

HUMBER'S RECORD OF MODERN ENGINEERING. SECOND SERIES.
Imp. 4to, with 36 Double Plates, Photographic Portrait of Robert Stephenson, C.E., M.P., F.R.S., &c., and copious descriptive Letterpress, Specifications, &c., £3 3s. half-morocco.

List of the Plates and Diagrams.

Birkenhead Docks, Low Water Basin (15 plates); Charing Cross Station Roof, C. C. Railway (3 plates); Digswell Viaduct, Great Northern Railway; Robbery Wood Viaduct, Great Northern Railway; Iron Permanent Way; Clydach Viaduct, Merthyr, Tredegar, and Abergavenny Railway; Ebbw Viaduct, Merthyr, Tredegar, and Abergavenny Railway; College Wood Viaduct, Cornwall Railway; Dublin Winter Palace Roof (3 plates); Bridge over the Thames, L. C. & D. Railway (6 plates); Albert Harbour, Greenock (4 plates).

"Mr. Humber has done the profession good and true service, by the fine collection of examples he has here brought before the profession and the public."—*Practical Mechanic's Journal.*

HUMBER'S RECORD OF MODERN ENGINEERING. THIRD SERIES.
Imp. 4to, with 40 Double Plates, Photographic Portrait of J. R. M'Clean, late Pres. Inst. C.E., and copious descriptive Letterpress, Specifications, &c., £3 3s. half-morocco.

List of the Plates and Diagrams.

MAIN DRAINAGE, METROPOLIS.—*North Side.*—Map showing Interception of Sewers; Middle Level Sewer (2 plates); Outfall Sewer, Bridge over River Lea (3 plates); Outfall Sewer, Bridge over Marsh Lane, North Woolwich Railway, and Bow and Barking Railway Junction; Outfall Sewer, Bridge over Bow and Barking Railway (3 plates); Outfall Sewer, Bridge over East London Waterworks' Feeder (2 plates); Outfall Sewer, Reservoir (2 plates); Outfall Sewer, Tumbling Bay and Outlet; Outfall Sewer, Penstocks. *South Side.*—Outfall Sewer, Bermondsey Branch (2 plates); Outfall Sewer, Reservoir and Outlet (4 plates); Outfall Sewer, Filth Hoist; Sections of Sewers (North and South Sides).
THAMES EMBANKMENT.—Section of River Wall; Steamboat Pier, Westminster (2 plates); Landing Stairs between Charing Cross and Waterloo Bridges; York Gate (2 plates); Overflow and Outlet at Savoy Street Sewer (3 plates); Steamboat Pier, Waterloo Bridge (3 plates); Junction of Sewers, Plans and Sections; Gullies, Plans and Sections; Rolling Stock; Granite and Iron Forts.

"The drawings have a constantly increasing value, and whoever desires to possess clear representations of the two great works carried out by our Metropolitan Board will obtain Mr. Humber's volume."—*Engineer.*

HUMBER'S RECORD OF MODERN ENGINEERING. FOURTH SERIES.
Imp. 4to, with 36 Double Plates, Photographic Portrait of John Fowler, late Pres. Inst. C.E., and copious descriptive Letterpress, Specifications, &c., £3 3s. half-morocco.

List of the Plates and Diagrams.

Abbey Mills Pumping Station, Main Drainage, Metropolis (4 plates); Barrow Docks (5 plates); Manquis Viaduct, Santiago and Valparaiso Railway (2 plates); Adam's Locomotive, St. Helen's Canal Railway (2 plates); Cannon Street Station Roof, Charing Cross Railway (3 plates); Road Bridge over the River Moka (2 plates); Telegraphic Apparatus for Mesopotamia; Viaduct over the River Wye, Midland Railway (3 plates); St. Germans Viaduct, Cornwall Railway (2 plates); Wrought-Iron Cylinder for Diving Bell; Millwall Docks (6 plates); Milroy's Patent Excavator; Metropolitan District Railway (6 plates); Harbours, Ports, and Breakwaters (3 plates).

"We gladly welcome another year's issue of this valuable publication from the able pen of Mr. Humber. The accuracy and general excellence of this work are well known, while its usefulness in giving the measurements and details of some of the latest examples of engineering, as carried out by the most eminent men in the profession, cannot be too highly prized."—*Artizan.*

CIVIL ENGINEERING, SURVEYING, etc. 9

MR. HUMBER'S ENGINEERING BOOKS—continued.

Strains, Calculation of.

A HANDY BOOK FOR THE CALCULATION OF STRAINS IN GIRDERS AND SIMILAR STRUCTURES, AND THEIR STRENGTH. Consisting of Formulæ and Corresponding Diagrams, with numerous details for Practical Application, &c. By WILLIAM HUMBER, A-M. Inst. C.E., &c. Fourth Edition. Crown 8vo, nearly 100 Woodcuts and 3 Plates, 7s. 6d. cloth.

"The formulæ are neatly expressed, and the diagrams good."—*Athenæum.*
"We heartily commend this really *handy* book to our engineer and architect readers."—*English Mechanic.*

Barlow's Strength of Materials, enlarged by Humber

A TREATISE ON THE STRENGTH OF MATERIALS: with Rules for Application in Architecture, the Construction of Suspension Bridges, Railways, &c. By PETER BARLOW, F.R.S. A New Edition, revised by his Sons, P. W. BARLOW, F.R.S., and W. H. BARLOW, F.R.S.; to which are added, Experiments by HODGKINSON, FAIRBAIRN, and KIRKALDY; and Formulæ for Calculating Girders, &c. Arranged and Edited by W. HUMBER, A-M. Inst. C.E. Demy 8vo, 400 pp., with 19 large Plates and numerous Woodcuts, 18s. cloth.

"Valuable alike to the student, tyro, and the experienced practitioner, it will always rank in future, as it has hitherto done, as the standard treatise on that particular subject."—*Engineer.*
"There is no greater authority than Barlow."—*Building News.*
"Deserves a foremost place on the bookshelves of every civil engineer."—*English Mechanic.*

Trigonometrical Surveying.

AN OUTLINE OF THE METHOD OF CONDUCTING A TRIGONOMETRICAL SURVEY, for the Formation of Geographical and Topographical Maps and Plans, Military Reconnaissance, Levelling, &c., with Useful Problems, Formulæ, and Tables. By Lieut.-General FROME, R.E. Fourth Edition, Revised and partly Re-written by Major General Sir CHARLES WARREN, G.C.M.G., R.E. With 19 Plates and 115 Woodcuts, royal 8vo, 16s. cloth.

"The simple fact that a fourth edition has been called for is the best testimony to its merits. No words of praise from us can strengthen the position so well and so steadily maintained by this work. Sir Charles Warren has revised the entire work, and made such additions as were necessary to bring every portion of the contents up to the present date."—*Broad Arrow.*

Oblique Bridges.

A PRACTICAL AND THEORETICAL ESSAY ON OBLIQUE BRIDGES. With 13 large Plates. By the late GEORGE WATSON BUCK, M.I.C.E. Third Edition, revised by his Son, J. H. WATSON BUCK, M.I.C.E.; and with the addition of Description to Diagrams for Facilitating the Construction of Oblique Bridges, by W. H. BARLOW, M.I.C.E. Royal 8vo, 12s. cloth.

"The standard text-book for all engineers regarding skew arches is Mr. Buck's treatise, and it would be impossible to consult a better."—*Engineer.*
"Mr. Buck's treatise is recognised as a standard text-book, and his treatment has divested the subject of many of the intricacies supposed to belong to it. As a guide to the engineer and architect, on a confessedly difficult subject, Mr. Buck's work is unsurpassed."—*Building News.*

Bridge Construction.

EXAMPLES OF BRIDGE AND VIADUCT CONSTRUCTION OF MASONRY, TIMBER, AND IRON. Consisting of 46 Plates from the Contract Drawings or Admeasurement of Select Works. By W. D. HASKOLL C.E. Second Edition, with the addition of 554 Estimates, and the Practice of Setting out Works. Illustrated with 6 pages of Diagrams. Imp. 4to, £2 12s. 6d. half-morocco.

"A work of the present nature by a man of Mr. Haskoll's experience must prove invaluable. The tables of estimates will considerably enhance its value."—*Engineering.*

Earthwork.

EARTHWORK TABLES. Showing the Contents in Cubic Yards of Embankments, Cuttings, &c., of Heights or Depths up to an average of 80 feet. By JOSEPH BROADBENT, C.E., and FRANCIS CAMPIN, C.E. Crown 8vo, 5s. cloth.

"The way in which accuracy is attained, by a simple division of each cross section into three elements, two in which are constant and one variable, is ingenious."—*Athenæum.*

Statics, Graphic and Analytic.

GRAPHIC AND ANALYTIC STATICS, in their Practical Application to the Treatment of Stresses in Roofs, Solid Girders, Lattice, Bowstring and Suspension Bridges, Braced Iron Arches and Piers, and other Frameworks. By R. HUDSON GRAHAM, C.E. Containing Diagrams and Plates to Scale. With numerous Examples, many taken from existing Structures. Specially arranged for Class-work in Colleges and Universities. Second Edition, Revised and Enlarged. 8vo, 16s. cloth.

"Mr. Graham's book will find a place wherever graphic and analytic statics are used or studied."—*Engineer.*
"This exhaustive treatise is admirably adapted for the architect and engineer, and will tend to wean the profession from a tedious and laboured mode of calculation."—*Building News.*
"The work is excellent from a practical point of view, and has evidently been prepared with much care. The directions for working are ample, and are illustrated by an abundance of well-selected examples. It is an excellent text-book for the practical draughtsman."—*Athenæum.*

Survey Practice.

AID TO SURVEY PRACTICE, for Reference in Surveying, Levelling, Setting-out and in Route Surveys of Travellers by Land and Sea. With Tables, Illustrations, and Records. By LOWIS D'A. JACKSON, A.M.I.C.E., Author of "Hydraulic Manual," "Modern Metrology," &c. Large crown 8vo, 12s. 6d. cloth.

"Mr. Jackson has produced a valuable *vade-mecum* for the surveyor. We can recommend this book as containing an admirable supplement to the teaching of the accomplished surveyor."—*Athenæum.*
"As a text-book we should advise all surveyors to place it in their libraries, and study well the matured instructions afforded in its pages."—*Colliery Guardian.*
"The author brings to his work a fortunate union of theory and practical experience which, aided by a clear and lucid style of writing, renders the book a very useful one."—*Builder.*

Surveying, Land and Marine.

LAND AND MARINE SURVEYING, in Reference to the Preparation of Plans for Roads and Railways; Canals, Rivers, Towns' Water Supplies; Docks and Harbours. With Description and Use of Surveying Instruments. By W. DAVIS HASKOLL, C.E., Author of "Bridge and Viaduct Construction," &c. Second Edition, Revised, with Additions. Large crown 8vo, 9s. cloth.

"A most useful and well arranged book for the aid of a student. We can strongly recommend it as a carefully written and valuable text-book. It enjoys a well-deserved repute among surveyors."—*Builder.*
"This volume cannot fail to prove of the utmost practical utility. It may be safely recommended to all students who aspire to become clean and expert surveyors."—*Mining Journal.*

Levelling.

A TREATISE ON THE PRINCIPLES AND PRACTICE OF LEVELLING. Showing its Application to purposes of Railway and Civil Engineering, in the Construction of Roads; with Mr. TELFORD's Rules for the same. By FREDERICK W. SIMMS, F.G.S., M. Inst. C.E. Seventh Edition, with the addition of LAW's Practical Examples for Setting-out Railway Curves, and TRAUTWINE's Field Practice of Laying-out Circular Curves. With 7 Plates and numerous Woodcuts, 8vo, 8s. 6d. cloth. *** TRAUTWINE on Curves may be had separate, 5s.

"The text-book on levelling in most of our engineering schools and colleges."—*Engineer.*
"The publishers have rendered a substantial service to the profession, especially to the younger members, by bringing out the present edition of Mr. Simms's useful work."—*Engineering.*

Tunnelling.

PRACTICAL TUNNELLING. Explaining in detail the Setting-out of the works, Shaft-sinking and Heading-driving, Ranging the Lines and Levelling underground, Sub-Excavating, Timbering, and the Construction of the Brickwork of Tunnels, with the amount of Labour required for, and the Cost of, the various portions of the work. By FREDERICK W. SIMMS, F.G.S., M. Inst. C.E. Third Edition, Revised and Extended by D. KINNEAR CLARK, M. Inst. C.E. Imperial 8vo, with 21 Folding Plates and numerous Wood Engravings, 30s. cloth.

"The estimation in which Mr. Simms's book has been held for over thirty years cannot be more truly expressed than in the words of the late Prof. Rankine:—'The best source of information on the subject of tunnels is Mr. F. W. Simms's work on Practical Tunnelling.'"—*Architect.*
"Mr. Clark has added immensely to the value of the book."—*Engineer.*
"The additional chapters by Mr. Clark, containing as they do numerous examples of modern practice, bring the book well up to date."—*Engineering.*

Heat, Expansion by.

EXPANSION OF STRUCTURES BY HEAT. By JOHN KEILY, C.E., late of the Indian Public Works and Victorian Railway Departments. Crown 8vo, 3s. 6d. cloth. [*Just published.*

SUMMARY OF CONTENTS.

Section I. FORMULAS AND DATA.
Section II. METAL BARS.
Section III. SIMPLE FRAMES.
Section IV. COMPLEX FRAMES AND PLATES.
Section V. THERMAL CONDUCTIVITY.
Section VI. MECHANICAL FORCE OF HEAT.
Section VII. WORK OF EXPANSION AND CONTRACTION.
Section VIII. SUSPENSION BRIDGES.
Section IX. MASONRY STRUCTURES.

"The aim the author has set before him, viz., to show the effects of heat upon metallic and other structures, is a laudable one, for this is a branch of physics upon which the engineer or architect can find but little reliable and comprehensive data in books."—*Builder.*

"Whoever is concerned to know the effect of changes of temperature on such structures as suspension bridges and the like, could not do better than consult Mr. Keily's valuable and handy exposition of the geometrical principles involved in these changes."—*Scotsman.*

Practical Mathematics.

MATHEMATICS FOR PRACTICAL MEN: Being a Commonplace Book of Pure and Mixed Mathematics. Designed chiefly for the use of Civil Engineers, Architects and Surveyors. By OLINTHUS GREGORY, LL.D., F.R.A.S., Enlarged by HENRY LAW, C.E. 4th Edition, carefully Revised by J. R. YOUNG, formerly Professor of Mathematics, Belfast College. With 13 Plates, 8vo, £1 1s., cloth.

"The engineer or architect will here find ready to his hand rules for solving nearly every mathematical difficulty that may arise in his practice. The rules are in all cases explained by means of examples, in which every step of the process is clearly worked out."—*Builder.*

"One of the most serviceable books for practical mechanics. . . . It is an instructive book for the student, and a text-book for him who, having once mastered the subjects it treats of, needs occasionally to refresh his memory upon them."—*Building News.*

Hydraulic Tables.

HYDRAULIC TABLES, CO-EFFICIENTS, and FORMULÆ *for finding the Discharge of Water from Orifices, Notches, Weirs, Pipes, and Rivers.* With New Formulæ, Tables, and General Information on Rainfall, Catchment-Basins, Drainage, Sewerage, Water Supply for Towns and Mill Power. By JOHN NEVILLE, Civil Engineer, M.R.I.A. Third Edition, carefully revised, with considerable Additions. Numerous Illustrations. Crown 8vo, 14s. cloth.

"It is, of all English books on the subject, the one nearest to completeness. . . . From the good arrangement of the matter, the clear explanations, and abundance of formulæ, the carefully calculated tables, and, above all, the thorough acquaintance with both theory and construction, which is displayed from first to last, the book will be found to be an acquisition."—*Architect.*

River Engineering.

RIVER BARS: *The Causes of their Formation, and their Treatment by "Induced Tidal Scour."* With a Description of the Successful Reduction by this Method of the Bar at Dublin. By I. J. MANN, Assist. Eng. to the Dublin Port and Docks Board. Royal 8vo, 7s. 6d. cloth.

"We recommend all interested in harbour works—and, indeed, those concerned in the improvements of rivers generally—to read Mr. Mann's interesting work."—*Engineer.*

"A most valuable contribution to the history of this branch of engineering."—*Engineering and Mining Journal.*

Hydraulics.

HYDRAULIC MANUAL. Consisting of Working Tables and Explanatory Text. Intended as a Guide in Hydraulic Calculations and Field Operations. By LOWIS D'A. JACKSON, Author of "Aid to Survey Practice," "Modern Metrology," &c. Fourth Edition. Rewritten and Enlarged. Large crown 8vo, 16s. cloth.

"The author has been a careful observer of the facts which have come under his notice, as well as a painstaking collector and critic of the results of the experiments of others, and from the great mass of material at his command he has constructed a manual which may be accepted as a trustworthy guide to this branch of the engineer's profession."—*Engineering.*

"The standard work in this department of mechanics. The present edition has been brought abreast of the most recent practice."—*Scotsman.*

"The most useful feature of this work is its freedom from what is superannuated, and its thorough adoption of recent experiments; the text is, in fact, in great part a short account of the great modern experiments."—*Nature.*

Drainage.

ON THE DRAINAGE OF LANDS, TOWNS AND BUILDINGS. By G. D. DEMPSEY, C.E., Author of "The Practical Railway Engineer," &c. Revised, with large Additions on RECENT PRACTICE IN DRAINAGE ENGINEERING, by D. KINNEAR CLARK, M.Inst. C.E. Author of "Tramways: their Construction and Working," "A Manual of Rules, Tables, and Data for Mechanical Engineers," &c. &c. Crown 8vo, 7s. 6d. cloth.
[*Just Published.*
"The new matter added to Mr. Dempsey's excellent work is characterised by the comprehensive grasp and accuracy of detail for which the name of Mr. D. K. Clark is a sufficient voucher."—*Athenæum.*
"As a work on recent practice in drainage engineering, the book is to be commended to all who are making that branch of engineering science their special study."—*Iron.*
'A comprehensive manual on drainage engineering, and a useful introduction to the student.'—*Building News.*

Tramways and their Working.

TRAMWAYS: THEIR CONSTRUCTION AND WORKING. Embracing a Comprehensive History of the System; with an exhaustive Analysis of the various Modes of Traction, including Horse-Power, Steam, Heated Water, and Compressed Air; a Description of the Varieties of Rolling Stock; and ample Details of Cost and Working Expenses: the Progress recently made in Tramway Construction, &c. &c. By D. KINNEAR CLARK, M. Inst. C.E. With over 200 Wood Engravings, and 13 Folding Plates. Two Vols., large crown 8vo, 30s. cloth.
"All interested in tramways must refer to it, as all railway engineers have turned to the author's work 'Railway Machinery.'"—*Engineer.*
"An exhaustive and practical work on tramways, in which the history of this kind of locomotion, and a description and cost of the various modes of laying tramways, are to be found."—*Building News.*
"The best form of rails, the best mode of construction, and the best mechanical appliances are so fairly indicated in the work under review, that any engineer about to construct a tramway will be enabled at once to obtain the practical information which will be of most service to him."—*Athenæum.*

Oblique Arches.

A PRACTICAL TREATISE ON THE CONSTRUCTION OF OBLIQUE ARCHES. By JOHN HART. Third Edition, with Plates. Imperial 8vo, 8s. cloth.

Curves, Tables for Setting-out.

TABLES OF TANGENTIAL ANGLES AND MULTIPLES *for Setting-out Curves from 5 to 200 Radius.* By ALEXANDER BEAZELEY, M. Inst. C.E. Third Edition. Printed on 48 Cards, and sold in a cloth box, waistcoat-pocket size, 3s. 6d.
"Each table is printed on a small card, which, being placed on the theodolite, leaves the hands free to manipulate the instrument—no small advantage as regards the rapidity of work."—*Engineer.*
"Very handy; a man may know that all his day's work must fall on two of these cards, which he puts into his own card-case, and leaves the rest behind."—*Athenæum.*

Engineering Fieldwork.

THE PRACTICE OF ENGINEERING FIELDWORK, *applied to Land and Hydraulic, Hydrographic, and Submarine Surveying and Levelling.* Second Edition, Revised, with considerable Additions, and a Supplement on Waterworks, Sewers, Sewage, and Irrigation. By W. DAVIS HASKOLL, C.E. Numerous Folding Plates. In One Volume, demy 8vo, £1 5s. cloth.

Tunnel Shafts.

THE CONSTRUCTION OF LARGE TUNNEL SHAFTS: A *Practical and Theoretical Essay.* By J. H. WATSON BUCK, M. Inst. C.E., Resident Engineer, London and North-Western Railway. Illustrated with Folding Plates, royal 8vo, 12s. cloth.
"Many of the methods given are of extreme practical value to the mason; and the observations on the form of arch, the rules for ordering the stone, and the construction of the templates will be found of considerable use. We commend the book to the engineering profession."—*Building News.*
"Will be regarded by civil engineers as of the utmost value, and calculated to save much time and obviate many mistakes."—*Colliery Guardian.*

Girders, Strength of.

GRAPHIC TABLE FOR FACILITATING THE COMPUTATION OF THE WEIGHTS OF WROUGHT IRON AND STEEL GIRDERS, *etc.,* for Parliamentary and other Estimates. By J. H. WATSON BUCK, M. Inst. C.E. On a Sheet, 2s. 6d.

Trusses.

TRUSSES OF WOOD AND IRON. *Practical Applications of Science in Determining the Stresses, Breaking Weights, Safe Loads, Scantlings, and Details of Construction,* with Complete Working Drawings. By WILLIAM GRIFFITHS, Surveyor, Assistant Master, Tranmere School of Science and Art. Oblong 8vo, 4s. 6d. cloth.

"This handy little book enters so minutely into every detail connected with the construction of roof trusses, that no student need be ignorant of these matters for want of an easy source to come at the knowledge."—*Practical Engineer.*

"From the manner of treating the subject, Mr. Griffiths' book is clear enough to enable a student to be his own teacher. It will be useful in the science school and in workshops."—*Architect.*

Railway Working.

SAFE RAILWAY WORKING. *A Treatise on Railway Accidents: Their Cause and Prevention; with a Description of Modern Appliances and Systems.* By CLEMENT E. STRETTON, C.E., Vice-President and Consulting Engineer, Amalgamated Society of Railway Servants. With Illustrations and Coloured Plates, crown 8vo, 4s. 6d. strongly bound.

OUTLINE OF CONTENTS.

Chapter I. SUMMARY OF ACCIDENTS, 1885.—II PERMANENT WAY.—III. SIGNALLING; BLOCK SYSTEM.—IV. CONTINUOUS BRAKES.—V. BREAKING OF RAILWAY AXLES.—VI. RAILWAY COUPLINGS.—VII. RAILWAY SERVANTS AND THE LAW OF MANSLAUGHTER.—Appendix I. RAILWAY TRAFFIC RETURNS.—II. RAILWAY SIGNAL RETURNS.—III. CONTINUOUS BRAKES RETURNS.—IV. MIXED TRAINS.

*** OPINIONS OF THE PRESS.

"A book for the engineer, the directors the managers; and, in short, all who wish for information on railway matters will find a perfect encyclopædia in 'Safe Railway Working.'"—*Railway Review.*

"Mr. Clement E. Stretton, the energetic Vice-President of the Amalgamated Society of Railway Servants, may be congratulated on having collected, in a very convenient form, much valuable information on the principal questions affecting the safe working of railways."—*Railway Engineer.*

"We commend the remarks on railway signalling to all railway managers, especially where a uniform code and practice is advocated."—*Herepath's Railway Journal.*

Field-Book for Engineers.

THE ENGINEER'S, MINING SURVEYOR'S, AND CONTRACTOR'S FIELD-BOOK. Consisting of a Series of Tables, with Rules, Explanations of Systems, and use of Theodolite for Traverse Surveying and Plotting the Work with minute accuracy by means of Straight Edge and Set Square only; Levelling with the Theodolite, Casting-out and Reducing Levels to Datum, and Plotting Sections in the ordinary manner; setting-out Curves with the Theodolite by Tangential Angles and Multiples, with Right and Left-hand Readings of the Instrument: Setting-out Curves without Theodolite, on the System of Tangential Angles by sets of Tangents and Offsets: and Earthwork Tables to 80 feet deep, calculated for every 6 inches in depth. By W. DAVIS HASKOLL, C.E. With numerous Woodcuts. Fourth Edition, Enlarged. Crown 8vo, 12s. cloth.

"The book is very handy, and the author might have added that the separate tables of sines and tangents to every minute will make it useful for many other purposes, the genuine traverse tables existing all the same."—*Athenæum.*

"Every person engaged in engineering field operations will estimate the importance of such a work and the amount of valuable time which will be saved by reference to a set of reliable tables prepared with the accuracy and fulness of those given in this volume."—*Railway News.*

Earthwork, Measurement of.

A MANUAL ON EARTHWORK. By ALEX. J. S. GRAHAM, C.E. With numerous Diagrams. 18mo, 2s. 6d. cloth.

"A great amount of practical information, very admirably arranged, and available for rough estimates, as well as for the more exact calculations required in the engineer's and contractor's offices."—*Artisan.*

Strains in Ironwork.

THE STRAINS ON STRUCTURES OF IRONWORK; with Practical Remarks on Iron Construction. By F. W. SHEILDS, M. Inst. C.E. Second Edition, with 5 Plates. Royal 8vo, 5s. cloth.

"The student cannot find a better little book on this subject."—*Engineer.*

Cast Iron and other Metals, Strength of.

A PRACTICAL ESSAY ON THE STRENGTH OF CAST IRON AND OTHER METALS. By THOMAS TREDGOLD, C.E. Fifth Edition, including HODGKINSON'S Experimental Researches. 8vo, 12s. cloth.

ARCHITECTURE, BUILDING, etc.

Construction.
THE SCIENCE OF BUILDING: An Elementary Treatise on the Principles of Construction. By E. WYNDHAM TARN, M.A., Architect. Second Edition, Revised, with 58 Engravings. Crown 8vo, 7s. 6d. cloth.

"A very valuable book, which we strongly recommend to all students."—*Builder.*

"No architectural student should be without this handbook of constructional knowledge."—*Architect.*

Villa Architecture.
A HANDY BOOK OF VILLA ARCHITECTURE: Being a Series of Designs for Villa Residences in various Styles. With Outline Specifications and Estimates. By C. WICKES, Architect, Author of "The Spires and Towers of England," &c. 61 Plates, 4to, £1 11s. 6d. half-morocco, gilt edges.

"The whole of the designs bear evidence of their being the work of an artistic architect, and they will prove very valuable and suggestive."—*Building News.*

Text-Book for Architects.
THE ARCHITECT'S GUIDE: Being a Text-Book of Useful Information for Architects, Engineers, Surveyors, Contractors, Clerks of Works, &c. &c. By FREDERICK ROGERS, Architect, Author of "Specifications for Practical Architecture," &c. Second Edition, Revised and Enlarged. With numerous Illustrations. Crown 8vo, 6s. cloth.

"As a text-book of useful information for architects, engineers, surveyors, &c., it would be hard to find a handier or more complete little volume."—*Standard.*

"A young architect could hardly have a better guide-book."—*Timber Trades Journal.*

Taylor and Cresy's Rome.
THE ARCHITECTURAL ANTIQUITIES OF ROME. By the late G. L. TAYLOR, Esq., F.R.I.B.A., and EDWARD CRESY, Esq. New Edition, thoroughly revised by the Rev. ALEXANDER TAYLOR, M.A. (son of the late G. L. Taylor, Esq.), Fellow of Queen's College, Oxford, and Chaplain of Gray's Inn. Large folio, with 130 Plates, half-bound, £3 3s.

N.B.—*This is the only book which gives on a large scale, and with the precision of architectural measurement, the principal Monuments of Ancient Rome in plan, elevation, and detail.*

"Taylor and Cresy's work has from its first publication been ranked among those professional books which cannot be bettered. . . . It would be difficult to find examples of drawings, even among those of the most painstaking students of Gothic, more thoroughly worked out than are the one hundred and thirty plates in this volume."—*Architect.*

Architectural Drawing.
PRACTICAL RULES ON DRAWING, for the Operative Builder and Young Student in Architecture. By GEORGE PYNE. With 14 Plates, 4to, 7s. 6d. boards.

Civil Architecture.
THE DECORATIVE PART OF CIVIL ARCHITECTURE. By Sir WILLIAM CHAMBERS, F.R.S. With Illustrations, Notes, and an Examination of Grecian Architecture, by JOSEPH GWILT, F.S.A. Edited by W. H. LEEDS. 66 Plates, 4to, 21s. cloth.

House Building and Repairing.
THE HOUSE-OWNER'S ESTIMATOR; or, What will it Cost to Build, Alter, or Repair? A Price Book adapted to the Use of Unprofessional People, as well as for the Architectural Surveyor and Builder. By the late JAMES D. SIMON, A.R.I.B.A. Edited and Revised by FRANCIS T. W. MILLER, A.R.I.B.A. With numerous Illustrations. Third Edition, Revised. Crown 8vo, 3s. 6d. cloth.

"In two years it will repay its cost a hundred times over"—*Field.*

"A very handy book."—*English Mechanic.*

Designing, Measuring, and Valuing.

THE STUDENT'S GUIDE to the PRACTICE of MEASURING AND VALUING ARTIFICERS' WORKS. Containing Directions for taking Dimensions, Abstracting the same, and bringing the Quantities into Bill, with Tables of Constants, and copious Memoranda for the Valuation of Labour and Materials in the respective Trades of Bricklayer and Slater, Carpenter and Joiner, Painter and Glazier, Paperhanger, &c. With 8 Plates and 63 Woodcuts. Originally edited by EDWARD DOBSON, Architect. Fifth Edition, Revised, with considerable Additions on Mensuration and Construction, and a New Chapter on Dilapidations, Repairs, and Contracts, by E. WYNDHAM TARN, M.A. Crown 8vo, 9s. cloth.

"Well fulfils the promise of its title-page, and we can thoroughly recommend it to the class for whose use it has been compiled. Mr. Tarn's additions and revisions have much increased the usefulness of the work, and have especially augmented its value to students."—*Engineering.*

"The work has been carefully revised and edited by Mr. E. Wyndham Tarn, M.A., and comprises several valuable additions on construction, mensuration, dilapidations and repairs, and other matters. . . . This edition will be found the most complete treatise on the principles of measuring and valuing artificers' work that has yet been published."—*Building News.*

Pocket Estimator and Technical Guide.

THE POCKET TECHNICAL GUIDE, MEASURER AND ESTIMATOR FOR BUILDERS AND SURVEYORS. Containing Technical Directions for Measuring Work in all the Building Trades, with a Treatise on the Measurement of Timber and Complete Specifications for Houses, Roads, and Drains, and an easy Method of Estimating the various parts of a Building collectively. By A. C. BEATON, Author of "Quantities and Measurements," &c. Fourth Edition, carefully Revised and Priced according to the Present Value of Materials and Labour, with 53 Woodcuts, leather, waistcoat-pocket size, 1s. 6d. gilt edges.

"No builder, architect, surveyor, or valuer should be without his 'Beaton's Guide.'"—*Building News.*

"Contains an extraordinary amount of information in daily requisition in measuring and estimating. Its presence in the pocket will save valuable time and trouble."—*Building World.*

"An exceedingly handy pocket companion, thoroughly reliable."—*Builder's Weekly Reporter.*

"This neat little compendium contains all that is requisite in carrying out contracts for excavating, tiling, bricklaying, paving, &c."—*British Trade Journal.*

Donaldson on Specifications.

THE HANDBOOK OF SPECIFICATIONS; or, Practical Guide to the Architect, Engineer, Surveyor, and Builder, in drawing up Specifications and Contracts for Works and Constructions. Illustrated by Precedents of Buildings actually executed by eminent Architects and Engineers. By Professor T. L. DONALDSON, P.R.I.B.A., &c. New Edition, in One large Vol., 8vo, with upwards of 1,000 pages of Text, and 33 Plates, £1 11s. 6d. cloth.

"In this work forty-four specifications of executed works are given, including the specifications for parts of the new Houses of Parliament, by Sir Charles Barry, and for the new Royal Exchange, by Mr. Tite, M.P. The latter, in particular, is a very complete and remarkable document. It embodies, to a great extent, as Mr. Donaldson mentions, 'the bill of quantities with the description of the works.' . . . It is valuable as a record, and more valuable still as a book of precedents. . . . Suffice it to say that Donaldson's 'Handbook of Specifications' must be bought by all architects."—*Builder.*

Bartholomew and Rogers' Specifications.

SPECIFICATIONS FOR PRACTICAL ARCHITECTURE. A Guide to the Architect, Engineer, Surveyor, and Builder. With an Essay on the Structure and Science of Modern Buildings. Upon the Basis of the Work by ALFRED BARTHOLOMEW, thoroughly Revised, Corrected, and greatly added to by FREDERICK ROGERS, Architect. Second Edition, Revised, with Additions. With numerous Illusts., medium 8vo, 15s. cloth.

"The collection of specifications prepared by Mr. Rogers on the basis of Bartholomew's work is too well known to need any recommendation from us. It is one of the books with which every young architect must be equipped ; for time has shown that the specifications cannot be set aside through any defect in them."—*Architect.*

"Good forms for specifications are of considerable value, and it was an excellent idea to compile a work on the subject upon the basis of the late Alfred Bartholomew's valuable work. The second edition of Mr. Rogers's book is evidence of the want of a book dealing with modern requirements and materials."—*Building News.*

Geometry for the Architect, Engineer, etc.

PRACTICAL GEOMETRY, for the Architect, Engineer and Mechanic. Giving Rules for the Delineation and Application of various Geometrical Lines, Figures and Curves. By E. W. TARN, M.A., Architect, Author of "The Science of Building," &c. Second Edition. With Appendices on Diagrams of Strains and Isometrical Projection. With 172 Illustrations, demy 8vo, 9s. cloth.

"No book with the same objects in view has ever been published in which the clearness of the rules laid down and the illustrative diagrams have been so satisfactory."—*Scotsman.*

"This is a manual for the practical man, whether architect, engineer, or mechanic. . . . The object of the author being to avoid all abstruse formulæ or complicated methods, and to enable persons with but a moderate knowledge of geometry to work out the problems required."—*English Mechanic.*

The Science of Geometry.

THE GEOMETRY OF COMPASSES; or, Problems Resolved by the mere Description of Circles, and the use of Coloured Diagrams and Symbols. By OLIVER BYRNE. Coloured Plates. Crown 8vo, 3s. 6d. cloth.

"The treatise is a good one, and remarkable—like all Mr. Byrne's contributions to the science of geometry—for the lucid character of its teaching."—*Building News.*

DECORATIVE ARTS, etc.

Woods and Marbles (Imitation of).

SCHOOL OF PAINTING FOR THE IMITATION OF WOODS AND MARBLES, as Taught and Practised by A. R. VAN DER BURG and P. VAN DER BURG, Directors of the Rotterdam Painting Institution. Royal folio, 18¾ by 12¼ in., Illustrated with 24 full-size Coloured Plates; also 12 plain Plates, comprising 154 Figures. Second and Cheaper Edition. Price £1 11s. 6d.

List of Plates.

1. Various Tools required for Wood Painting —2, 3. Walnut: Preliminary Stages of Graining and Finished Specimen — 4. Tools used for Marble Painting and Method of Manipulation— 5, 6. St. Remi Marble: Earlier Operations and Finished Specimen—7. Methods of Sketching different Grains, Knots, &c.—8, 9. Ash: Preliminary Stages and Finished Specimen — 10. Methods of Sketching Marble Grains—11, 12. Breche Marble: Preliminary Stages of Working and Finished Specimen—13. Maple: Methods of Producing the different Grains—14, 15. Bird's-eye Maple: Preliminary Stages and Finished Specimen—16. Methods of Sketching the different Species of White Marble—17, 18. White Marble: Preliminary Stages of Process and Finished Specimen—19. Mahogany: Specimens of various Grains and Methods of Manipulation—20, 21. Mahogany: Earlier Stages and Finished Specimen—22, 23, 24. Sienna Marble: Varieties of Grain, Preliminary Stages and Finished Specimen—25, 26, 27. Juniper Wood: Methods of producing Grain, &c.; Preliminary Stages and Finished Specimen—28, 29, 30. Vert de Mer Marble: Varieties of Grain and Methods of Working Unfinished and Finished Specimens—31, 32, 33. Oak: Varieties of Grain, Tools Employed, and Methods of Manipulation, Preliminary Stages and Finished Specimen—34, 35, 36. Waulsort Marble: Varieties of Grain, Unfinished and Finished Specimens.

*** OPINIONS OF THE PRESS.

"Those who desire to attain skill in the art of painting woods and marbles will find advantage in consulting this book. . . . Some of the Working Men's Clubs should give their young men the opportunity to study it."—*Builder.*

"A comprehensive guide to the art. The explanations of the processes, the manipulation and management of the colours, and the beautifully executed plates will not be the least valuable to the student who aims at making his work a faithful transcript of nature."—*Building News.*

"Students and novices are fortunate who are able to become the possessors of *so noble a work.*"—*Architect.*

House Decoration.

ELEMENTARY DECORATION. A Guide to the Simpler Forms of Everyday Art, as applied to the Interior and Exterior Decoration of Dwelling Houses, &c. By JAMES W. FACEY, Jun. With 68 Cuts. 12mo, 2s. cloth limp.

"As a technical guide-book to the decorative painter it will be found reliable."—*Building News.*

PRACTICAL HOUSE DECORATION: A Guide to the Art of Ornamental Painting, the Arrangement of Colours in Apartments, and the principles of Decorative Design. With some Remarks upon the Nature and Properties of Pigments. By JAMES WILLIAM FACEY, Author of "Elementary Decoration," &c. With numerous Illustrations. 12mo, 2s. 6d. cloth limp.

N.B.—The above Two Works together in One Vol., strongly half-bound, 5s.

Colour.

A GRAMMAR OF COLOURING. Applied to Decorative Painting and the Arts. By GEORGE FIELD. New Edition, Revised, Enlarged, and adapted to the use of the Ornamental Painter and Designer. By ELLIS A. DAVIDSON. With New Coloured Diagrams and Engravings. 12mo, 3s. 6d. cloth boards.

"The book is a most useful *resume* of the properties of pigments."—*Builder.*

House Painting, Graining, etc.

HOUSE PAINTING, GRAINING, MARBLING, AND SIGN WRITING, A Practical Manual of. By ELLIS A. DAVIDSON. Fifth Edition. With Coloured Plates and Wood Engravings. 12mo, 6s. cloth boards.

"A mass of information, of use to the amateur and of value to the practical man."—*English Mechanic.*

"Simply invaluable to the youngster entering upon this particular calling, and highly serviceable to the man who is practising it."—*Furniture Gazette.*

Decorators, Receipts for.

THE DECORATOR'S ASSISTANT: A Modern Guide to Decorative Artists and Amateurs, Painters, Writers, Gilders, &c. Containing upwards of 600 Receipts, Rules and Instructions; with a variety of Information for General Work connected with every Class of Interior and Exterior Decorations, &c. 152 pp., crown 8vo, 1s. in wrapper.

"Full of receipts of value to decorators, painters, gilders, &c. The book contains the gist of larger treatises on colour and technical processes. It would be difficult to meet with a work so full of varied information on the painter's art."—*Building News.*

"We recommend the work to all who, whether for pleasure or profit, require a guide to decoration."—*Plumber and Decorator.*

Moyr Smith on Interior Decoration.

ORNAMENTAL INTERIORS, ANCIENT AND MODERN. By J. MOYR SMITH. Super-royal 8vo, with 32 full-page Plates and numerous smaller Illusts., handsomely bound in cloth, gilt top, price 18s. [*Just published.*

☞ In "ORNAMENTAL INTERIORS" *the designs of more than thirty artist-decorators and architects of high standing have been illustrated. The book may therefore fairly claim to give a good general view of the works of the modern school of decoration, besides giving characteristic examples of earlier decorative arrangements.*

"ORNAMENTAL INTERIORS" *gives a short account of the styles of Interior Decoration as practised by the Ancients in Egypt, Greece, Assyria, Rome and Byzantium. This part is illustrated by characteristic designs. The main body of the work, however, is devoted to the illustration of the modern styles of Decorative Art, and many examples are given of decorative designs suitable for modern Dining-Rooms, Drawing-Rooms, Libraries, Staircases and Halls, Parlours, Studies and Smoking-Rooms. The Decoration of Public Buildings is illustrated by views of the chief State Apartments in Buckingham Palace and Windsor Castle, the Salle de Leys at Antwerp, the Salle de Mariages at Brussels, and of other works which have distinctive features suitable for the purpose.*

*** OPINIONS OF THE PRESS.

"The book is well illustrated and handsomely got up, and contains some true criticism and a good many good examples of decorative treatment."—*The Builder.*

"We can greatly commend Mr. Moyr Smith's book, for it is the production of one professedly capable in decorative work, and abounds with useful hints and descriptions of executed modern work, together with a well-put *resume* of ancient styles. . . . As much a book for the drawing-room as for the manufacturer."—*The British Architect.*

"Well fitted for the dilettante, amateur, and professional designer."—*Decoration.*

"This is the most elaborate, and beautiful work on the artistic decoration of interiors that we have seen. . . . The scrolls, panels and other designs from the author's own pen are very beautiful and chaste; but he takes care that the designs of other men shall figure even more than his own."—*Liverpool Albion.*

"To all who take an interest in elaborate domestic ornament this handsome volume will be welcome."—*Graphic.*

"Mr. Moyr Smith deserves the thanks of art workers for having placed within their reach a book that seems eminently adapted to afford, by example and precept, that guidance of which most craftsmen stand in need."—*Furniture Gazette.*

British and Foreign Marbles.

MARBLE DECORATION and *the Terminology of British and Foreign Marbles.* A Handbook for Students. By GEORGE H. BLAGROVE, Author of "Shoring and its Application," &c. With 28 Illustrations. Crown 8vo, 3s. 6d. cloth. [*Just published.*

DELAMOTTE'S WORKS ON ILLUMINATION AND ALPHABETS.

A PRIMER OF THE ART OF ILLUMINATION, for the Use of Beginners: with a Rudimentary Treatise on the Art, Practical Directions for its exercise, and Examples taken from Illuminated MSS., printed in Gold and Colours. By F. DELAMOTTE. New and Cheaper Edition. Small 4to, 6s. ornamental boards.

"The examples of ancient MSS. recommended to the student, which, with much good sense, the author chooses from collections accessible to all, are selected with judgment and knowledge, as well as taste."—*Athenæum.*

ORNAMENTAL ALPHABETS, Ancient and Mediæval, from the Eighth Century, with Numerals; including Gothic, Church-Text, large and small, German, Italian, Arabesque, Initials for Illumination, Monograms, Crosses, &c. &c., for the use of Architectural and Engineering Draughtsmen, Missal Painters, Masons, Decorative Painters, Lithographers, Engravers, Carvers, &c. &c. Collected and Engraved by F. DELAMOTTE, and printed in Colours. New and Cheaper Edition. Royal 8vo, oblong, 2s. 6d. ornamental boards.

"For those who insert enamelled sentences round gilded chalices, who blazon shop legends over shop-doors, who letter church walls with pithy sentences from the Decalogue, this book will be useful."—*Athenæum.*

EXAMPLES OF MODERN ALPHABETS, Plain and Ornamental; including German, Old English, Saxon, Italic, Perspective, Greek, Hebrew, Court Hand, Engrossing, Tuscan, Riband, Gothic, Rustic, and Arabesque; with several Original Designs, and an Analysis of the Roman and Old English Alphabets, large and small, and Numerals, for the use of Draughtsmen, Surveyors, Masons, Decorative Painters, Lithographers, Engravers, Carvers, &c. Collected and Engraved by F. DELAMOTTE, and printed in Colours. New and Cheaper Edition. Royal 8vo, oblong, 2s. 6d. ornamental boards.

"There is comprised in it every possible shape into which the letters of the alphabet and numerals can be formed, and the talent which has been expended in the conception of the various plain and ornamental letters is wonderful."—*Standard.*

MEDIÆVAL ALPHABETS AND INITIALS FOR ILLUMINATORS. By F. G. DELAMOTTE. Containing 21 Plates and Illuminated Title, printed in Gold and Colours. With an Introduction by J. WILLIS BROOKS. Fourth and Cheaper Edition. Small 4to, 4s. ornamental boards.

"A volume in which the letters of the alphabet come forth glorified in gilding and all the colours of the prism interwoven and intertwined and intermingled."—*Sun.*

THE EMBROIDERER'S BOOK OF DESIGN. Containing Initials, Emblems, Cyphers, Monograms, Ornamental Borders, Ecclesiastical Devices, Mediæval and Modern Alphabets, and National Emblems. Collected by F. DELAMOTTE, and printed in Colours. Oblong royal 8vo, 1s. 6d. ornamental wrapper.

"The book will be of great assistance to ladies and young children who are endowed with the art of plying the needle in this most ornamental and useful pretty work."—*East Anglian Times.*

Wood Carving.

INSTRUCTIONS IN WOOD-CARVING, for Amateurs; with Hints on Design. By A LADY. With Ten large Plates, 2s. 6d. in emblematic wrapper.

"The handicraft of the wood-carver, so well as a book can impart it, may be learnt from 'A Lady's' publication."—*Athenæum.*
"The directions given are plain and easily understood."—*English Mechanic.*

Glass Painting.

GLASS STAINING AND THE ART OF PAINTING ON GLASS. From the German of Dr. GESSERT and EMANUEL OTTO FROMBERG. With an Appendix on THE ART OF ENAMELLING. 12mo, 2s. 6d. cloth limp.

Letter Painting.

THE ART OF LETTER PAINTING MADE EASY. By JAMES GREIG BADENOCH. With 12 full-page Engravings of Examples, 1s. cloth limp.

"The system is a simple one, but quite original, and well worth the careful attention of letter painters. It can be easily mastered and remembered."—*Building News.*

CARPENTRY, TIMBER, etc.

Tredgold's Carpentry, Enlarged by Tarn.

THE ELEMENTARY PRINCIPLES OF CARPENTRY. A Treatise on the Pressure and Equilibrium of Timber Framing, the Resistance of Timber, and the Construction of Floors, Arches, Bridges, Roofs, Uniting Iron and Stone with Timber, &c. To which is added an Essay on the Nature and Properties of Timber, &c., with Descriptions of the kinds of Wood used in Building; also numerous Tables of the Scantlings of Timber for different purposes, the Specific Gravities of Materials, &c. By THOMAS TREDGOLD, C.E. With an Appendix of Specimens of Various Roofs of Iron and Stone, Illustrated. Seventh Edition, thoroughly revised and considerably enlarged by E. WYNDHAM TARN, M.A., Author of "The Science of Building," &c. With 61 Plates, Portrait of the Author, and several Woodcuts. In one large vol., 4to, price £1 5s. cloth.

"Ought to be in every architect's and every builder's library."—*Builder.*
"A work whose monumental excellence must commend it wherever skilful carpentry is concerned. The author's principles are rather confirmed than impaired by time. The additional plates are of great intrinsic value."—*Building News.*

Woodworking Machinery.

WOODWORKING MACHINERY: *Its Rise, Progress, and Construction.* With Hints on the Management of Saw Mills and the Economical Conversion of Timber. Illustrated with Examples of Recent Designs by leading English, French, and American Engineers. By M. POWIS BALE, A.M. Inst. C.E., M.I.M.E. Large crown 8vo, 12s. 6d. cloth.

"Mr. Bale is evidently an expert on the subject and he has collected so much information that his book is all-sufficient for builders and others engaged in the conversion of timber."—*Architect.*
"The most comprehensive compendium of wood-working machinery we have seen. The author is a thorough master of his subject."—*Building News.*
"The appearance of this book at the present time will, we should think, give a considerable impetus to the onward march of the machinist engaged in the designing and manufacture of wood-working machines. It should be in the office of every wood-working factory."—*English Mechanic.*

Saw Mills.

SAW MILLS: *Their Arrangement and Management, and the Economical Conversion of Timber.* (Being a Companion Volume to "Woodworking Machinery.") By M. POWIS BALE, A.M. Inst. C.E., M.I.M.E. With numerous Illustrations. Crown 8vo, 10s. 6d. cloth.

"The *administration* of a large sawing establishment is discussed, and the subject examined from a financial standpoint. Hence the size, shape, order, and disposition of saw-mills and the like are gone into in detail, and the course of the timber is traced from its reception to its delivery in its converted state. We could not desire a more complete or practical treatise."—*Builder.*
"We highly recommend Mr. Bale's work to the attention and perusal of all those who are engaged in the art of wood conversion, or who are about building or remodelling saw-mills on improved principles."—*Building News.*

Carpentering.

THE CARPENTER'S NEW GUIDE; or, Book of Lines for Carpenters; comprising all the Elementary Principles essential for acquiring a knowledge of Carpentry. Founded on the late PETER NICHOLSON'S Standard Work. A New Edition, revised by ARTHUR ASHPITEL, F.S.A. Together with Practical Rules on Drawing, by GEORGE PYNE. With 74 Plates, 4to. £1 1s. cloth.

Handrailing.

A PRACTICAL TREATISE ON HANDRAILING: Showing New and Simple Methods for Finding the Pitch of the Plank, Drawing the Moulds, Bevelling, Jointing-up, and Squaring the Wreath. By GEORGE COLLINGS. Illustrated with Plates and Diagrams. 12mo, 1s. 6d. cloth limp.

"Will be found of practical utility in the execution of this difficult branch of joinery."—*Builder*
"Almost every difficult phase of this somewhat intricate branch of joinery is elucidated by the aid of plates and explanatory letterpress."—*Furniture Gazette.*

Circular Work.

CIRCULAR WORK IN CARPENTRY AND JOINERY: A Practical Treatise on Circular Work of Single and Double Curvature. By GEORGE COLLINGS, Author of "A Practical Treatise on Handrailing." Illustrated with numerous Diagrams. 12mo, 2s. 6d. cloth limp. [*Just published.*

"An excellent example of what a book of this kind should be. Cheap in price, clear in definition and practical in the examples selected."—*Builder.*

Timber Merchant's Companion.

THE TIMBER MERCHANT'S AND BUILDER'S COMPANION. Containing New and Copious Tables of the Reduced Weight and Measurement of Deals and Battens, of all sizes, from One to a Thousand Pieces, and the relative Price that each size bears per Lineal Foot to any given Price per Petersburg Standard Hundred; the Price per Cube Foot of Square Timber to any given Price per Load of 50 Feet; the proportionate Value of Deals and Battens by the Standard, to Square Timber by the Load of 50 Feet; the readiest mode of ascertaining the Price of Scantling per Lineal Foot of any size, to any given Figure per Cube Foot, &c. &c. By WILLIAM DOWSING. Fourth Edition, Revised and Corrected. Cr. 8vo, 3s. cl.

"Every timber merchant and builder ought to possess it."—*Hull Advertiser.*
"We are glad to see a fourth edition of these admirable tables, which for correctness and simplicity of arrangement leave nothing to be desired."—*Timber Trades Journal.*
"An exceedingly well-arranged, clear, and concise manual of tables for the use of all who buy or sell timber."—*Journal of Forestry.*

Practical Timber Merchant.

THE PRACTICAL TIMBER MERCHANT. Being a Guide for the use of Building Contractors, Surveyors, Builders, &c., comprising useful Tables for all purposes connected with the Timber Trade, Marks of Wood, Essay on the Strength of Timber, Remarks on the Growth of Timber, &c. By W. RICHARDSON. Fcap. 8vo, 3s. 6d. cloth.

"This handy manual contains much valuable information for the use of timber merchants, builders, foresters, and all others connected with the growth, sale, and manufacture of timber."—*Journal of Forestry.*

Timber Freight Book.

THE TIMBER MERCHANT'S, SAW MILLER'S, AND IMPORTER'S FREIGHT BOOK AND ASSISTANT. Comprising Rules, Tables, and Memoranda relating to the Timber Trade. By WILLIAM RICHARDSON Timber Broker; together with a Chapter on "SPEEDS OF SAW MILL MACHINERY," by M. POWIS BALE, M.I.M.E. &c. 12mo, 3s. 6d. cl. boards.

"A very useful manual of rules, tables, and memoranda, relating to the timber trade. We recommend it as a compendium of calculation to all timber measurers and merchants, and as supplying a real want in the trade."—*Building News.*

Packing-Case Makers, Tables for.

PACKING-CASE TABLES; showing the number of Superficial Feet in Boxes or Packing-Cases, from six inches square and upwards. By W. RICHARDSON, Timber Broker. Second Edition. Oblong 4to, 3s. 6d. cl.

"Invaluable labour-saving tables."—*Ironmonger.* "Will save much labour."—*Grocer.*

Superficial Measurement.

THE TRADESMAN'S GUIDE TO SUPERFICIAL MEASUREMENT. Tables calculated from 1 to 200 inches in length, by 1 to 108 inches in breadth. For the use of Architects, Surveyors, Engineers, Timber Merchants, Builders, &c. By JAMES HAWKINGS. Third Edition. Fcap., 3s. 6d. cloth.

"A useful collection of tables to facilitate rapid calculation of surfaces. The exact area of any surface of which the limits have been ascertained can be instantly determined. The book will be found of the greatest utility to all engaged in building operations."—*Scotsman.*

Forestry.

THE ELEMENTS OF FORESTRY. Designed to afford Information concerning the Planting and Care of Forest Trees for Ornament or Profit, with Suggestions upon the Creation and Care of Woodlands. By F. B. HOUGH. Large crown 8vo, 10s. cloth.

Timber Importer's Guide.

THE TIMBER IMPORTER'S, TIMBER MERCHANT'S AND BUILDER'S STANDARD GUIDE. By RICHARD E. GRANDY. Comprising an Analysis of Deal Standards, Home and Foreign, with Comparative Values and Tabular Arrangements for fixing Nett Landed Cost on Baltic and North American Deals, including all intermediate Expenses, Freight, Insurance, &c. &c. Together with copious Information for the Retailer and Builder. Third Edition, Revised. 12mo, 2s. 6d. cloth boards.

"Everything it pretends to be: built up gradually, it leads one from a forest to a treenail, and throws in, as a makeweight, a host of material concerning bricks, columns, cisterns, &c."—*English Mechanic.*

NAVAL ARCHITECTURE, NAVIGATION, etc.

Chain Cables.
CHAIN CABLES AND CHAINS. Comprising Sizes and Curves of Links, Studs, &c., Iron for Cables and Chains, Chain Cable and Chain Making, Forming and Welding Links, Strength of Cables and Chains, Certificates for Cables, Marking Cables, Prices of Chain Cables and Chains, Historical Notes, Acts of Parliament, Statutory Tests, Charges for Testing, List of Manufacturers of Cables, &c. &c. By THOMAS W. TRAILL, F.E.R.N., M. Inst. C.E., Engineer Surveyor in Chief, Board of Trade, the Inspector of Chain Cable and Anchor Proving Establishments, and General Superintendent, Lloyd's Committee on Proving Establishments. With numerous Tables, Illustrations and Lithographic Drawings. Folio, £2 2s. cloth, bevelled boards.

"The author writes not only with a full acquaintance with scientific formulæ and details, but also with a profound and fully-instructed sense of the importance to the safety of our ships and sailors of fidelity in the manufacture of cables."—*Athenæum*.

"The business of chain cable making is well explained and illustrated. We can safely recommend this work to all in any way connected with the manufacture of chain cables and chains, as a good book."—*Nature*.

"It contains a vast amount of valuable information. Nothing seems to be wanting to make it a complete and standard work of reference on the subject."—*Nautical Magazine*.

Pocket-Book for Naval Architects and Shipbuilders.
THE NAVAL ARCHITECT'S AND SHIPBUILDER'S POCKET-BOOK of Formulæ, Rules, and Tables, and MARINE ENGINEER'S AND SURVEYOR'S Handy Book of Reference. By CLEMENT MACKROW, Member of the Institution of Naval Architects, Naval Draughtsman. Third Edition, Revised. With numerous Diagrams, &c. Fcap., 12s. 6d. strongly bound in leather.

"Should be used by all who are engaged in the construction or design of vessels. . . . Will be found to contain the most useful tables and formulæ required by shipbuilders, carefully collected from the best authorities, and put together in a popular and simple form."—*Engineer*.

"The professional shipbuilder has now, in a convenient and accessible form, reliable data for solving many of the numerous problems that present themselves in the course of his work."—*Iron*.

"There is scarcely a subject on which a naval architect or shipbuilder can require to refresh his memory which will not be found within the covers of Mr. Mackrow's book."—*English Mechanic*.

Pocket-Book for Marine Engineers.
A POCKET-BOOK OF USEFUL TABLES AND FORMULÆ FOR MARINE ENGINEERS. By FRANK PROCTOR, A.I.N.A. Third Edition. Royal 32mo, leather, gilt edges, with strap, 4s.

"We recommend it to our readers as going far to supply a long-felt want."—*Naval Science*.

"A most useful companion to all marine engineers."—*United Service Gazette*.

Lighthouses.
EUROPEAN LIGHTHOUSE SYSTEMS. Being a Report of a Tour of Inspection made in 1873. By Major GEORGE H. ELLIOT, Corps of Engineers, U.S.A. Illustrated by 51 Engravings and 31 Woodcuts. 8vo, 21s. cloth.

** The following are published in WEALE's RUDIMENTARY SERIES.

MASTING, MAST-MAKING, AND RIGGING OF SHIPS. By ROBERT KIPPING, N.A. Fifteenth Edition. 12mo, 2s. 6d. cloth boards.

SAILS AND SAIL-MAKING. Eleventh Edition, Enlarged, with an Appendix. By ROBERT KIPPING, N.A. Illustrated. 12mo, 3s. cloth boards.

NAVAL ARCHITECTURE. By JAMES PEAKE. Fifth Edition, with Plates and Diagrams. 12mo, 4s. cloth boards.

MARINE ENGINES AND STEAM VESSELS (*A Treatise on*). By ROBERT MURRAY, C.E., Principal Officer to the Board of Trade for the East Coast of Scotland District. Eighth Edition, thoroughly Revised, with considerable Additions, by the Author and by GEORGE CARLISLE, C.E., Senior Surveyor to the Board of Trade at Liverpool. 12mo, 5s. cloth boards.

PRACTICAL NAVIGATION. Consisting of THE SAILOR'S SEA-BOOK, by JAS. GREENWOOD and W. H. ROSSER; together with the requisite Mathematical and Nautical Tables for the Working of the Problems, by HENRY LAW, C.E. and Prof. J. R. YOUNG. Illustrated 12mo, 7s. half-bound.

MINING AND MINING INDUSTRIES.

Metalliferous Mining.

BRITISH MINING: A Treatise on the History, Discovery, Practical Development, and Future Prospects of Metalliferous Mines in the United Kingdom. By ROBERT HUNT, F.R.S., Keeper of Mining Records; Editor of "Ure's Dictionary of Arts, Manufactures, and Mines," &c. Upwards of 950 pp., with 230 Illustrations. Second Edition, Revised. Super-royal 8vo, £2 2s. cloth. *[Just published.*

** OPINIONS OF THE PRESS.

"One of the most valuable works of reference of modern times. Mr. Hunt, as keeper of mining records of the United Kingdom, has had opportunities for such a task not enjoyed by anyone else, and has evidently made the most of them. . . . The language and style adopted are good, and the treatment of the various subjects laborious, conscientious, and scientific."—*Engineering.*

"Probably no one in this country was better qualified than Mr. Hunt for undertaking such a work. Brought into frequent and close association during a long life-time with the principal guardians of our mineral and metallurgical industries, he enjoyed a position exceptionally favourable for collecting the necessary information. The use which he has made of his opportunities is sufficiently attested by the dense mass of information crowded into the handsome volume which has just been published. . . . In placing before the reader a sketch of the present position of British Mining, Mr. Hunt treats his subject so fully and illustrates it so amply that this section really forms a little treatise on practical mining. . . . The book is, in fact, a treasure-house of statistical information on mining subjects, and we know of no other work embodying so great a mass of matter of this kind. Were this the only merit of Mr. Hunt's volume it would be sufficient to render it indispensable in the library of everyone interested in the development of the mining and metallurgical industries of this country."—*Athenæum.*

"A mass of information not elsewhere available, and of the greatest value to those who may be interested in our great mineral industries."—*Engineer.*

"A sound, business-like collection of interesting facts. . . . The amount of information Mr. Hunt has brought together is enormous. . . . The volume appears likely to convey more instruction upon the subject than any work hitherto published."—*Mining Journal.*

"The work will be for the mining industry what Dr. Percy's celebrated treatise has been for the metallurgical—a book that cannot with advantage be omitted from the library."—*Iron and Coal Trades Review.*

"The literature of mining has hitherto possessed no work approaching in importance to that which has just been published. There is much in Mr. Hunt's valuable work that every shareholder in a mine should read with close attention. The entire subject of practical mining—from the first search for the lode to the latest stages of dressing the ore—is dealt with in a masterly manner."—*Academy.*

Coal and Iron.

THE COAL AND IRON INDUSTRIES OF THE UNITED KINGDOM. Comprising a Description of the Coal Fields, and of the Principal Seams of Coal, with Returns of their Produce and its Distribution, and Analyses of Special Varieties. Also an Account of the occurrence of Iron Ores in Veins or Seams; Analyses of each Variety; and a History of the Rise and Progress of Pig Iron Manufacture since the year 1740, exhibiting the Economies introduced in the Blast Furnaces for its Production and Improvement. By RICHARD MEADE, Assistant Keeper of Mining Records. With Maps of the Coal Fields and Ironstone Deposits of the United Kingdom. 8vo, £1 8s. cloth.

"The book is one which must find a place on the shelves of all interested in coal and iron production, and in the iron, steel, and other metallurgical industries."—*Engineer.*

"Of this book we may unreservedly say that it is the best of its class which we have ever met. . . . A book of reference which no one engaged in the iron or coal trades should omit from his library."—*Iron and Coal Trades Review.*

"An exhaustive treatise and a valuable work of reference."—*Mining Journal.*

Prospecting for Gold and other Metals.

THE PROSPECTOR'S HANDBOOK: A Guide for the Prospector and Traveller in Search of Metal-Bearing or other Valuable Minerals. By J. W. ANDERSON, M.A. (Camb.), F.R.G.S., Author of "Fiji and New Caledonia." Third Edition, Revised, with Additions. Small crown 8vo, 3s. 6d. cloth. *[Just published.*

"Will supply a much felt want, especially among Colonists, in whose way are so often thrown many mineralogical specimens the value of which it is difficult for anyone, not a specialist, to determine. The author has placed his instructions before his readers in the plainest possible terms, and his book is the best of its kind."—*Engineer.*

"How to find commercial minerals, and how to identify them when they are found, are the leading points to which attention is directed. The author has managed to pack as much practical detail into his pages as would supply material for a book three times its size."—*Mining Journal.*

"Those toilers who explore the trodden or untrodden tracks on the face of the globe will find much that is useful to them in this book."—*Athenæum.*

Mining Notes and Formulæ.

NOTES AND FORMULÆ FOR MINING STUDENTS. By JOHN HERMAN MERIVALE, M.A., Certificated Colliery Manager, Professor of Mining in the Durham College of Science, Newcastle-upon-Tyne. Second Edition, carefully Revised. Small crown 8vo, cloth, price 2s. 6d.
[*Just published.*

☞ *This book consists of a collection of notes and formulæ drawn from various sources, the authority being quoted in most instances. It is hoped that the work will be useful not only to students but to the profession.*

The principal sources of information upon mining matters are the Transactions of the various Engineering Societies to which the student, in most of our large towns, has access. A great many references to the most familiar of them are given, so that the student who wishes to follow up a subject may be in a position to acquaint himself with details which could not be included in a work like this.

The examples of the use of the formulæ, at the end of the book, are merely given to assist students working without a teacher.

" Invaluable to anyone who is working up for an examination on mining subjects."—*Coal and Iron Trades Review.*

" The author has done his work in an exceedingly creditable manner, and has produced a book that will be of service to students, and those who are practically engaged in mining operations."—*Engineer.*

" A vast amount of technical matter of the utmost value to mining engineers, and of considerable interest to students."—*Schoolmaster.*

Mineral Surveying and Valuing.

THE MINERAL SURVEYOR AND VALUER'S COMPLETE GUIDE, comprising a Treatise on Improved Mining Surveying and the Valuation of Mining Properties, with New Traverse Tables. By WM. LINTERN, Mining and Civil Engineer. Second Edition, with an Appendix on " Magnetic and Angular Surveying," with Records of the Peculiarities of Needle Disturbances. With Four Plates of Diagrams, Plans, &c. 12mo, 4s. cloth.
[*Just published.*

" An enormous fund of information of great value."—*Mining Journal.*

" Mr. Lintern's book forms a valuable and thoroughly trustworthy guide."—*Iron and Coal Trades Review.*

" This new edition must be of the highest value to colliery surveyors, proprietors and managers."—*Colliery Guardian.*

Metalliferous Minerals and Mining.

TREATISE ON METALLIFEROUS MINERALS AND MINING. By D. C. DAVIES, F.G.S., Mining Engineer, &c., Author of " A Treatise on Slate and Slate Quarrying." Illustrated with numerous Wood Engravings. Fourth Edition. Crown 8vo, 12s. 6d. cloth.

" Neither the practical miner nor the general reader interested in mines, can have a better book or his companion and his guide."—*Mining Journal.*

" The volume is one which no student of mineralogy should be without."—*Colliery Guardian.*

" We are doing our readers a service in calling their attention to this valuable work."—*Mining World.*

" A book that will not only be useful to the geologist, the practical miner, and the metallurgist, but also very interesting to the general public."—*Iron.*

" As a history of the present state of mining throughout the world this book has a real value, and it supplies an actual want, for no such information has hitherto been brought together within such limited space."—*Athenæum.*

Earthy Minerals and Mining.

A TREATISE ON EARTHY AND OTHER MINERALS AND MINING. By D. C. DAVIES, F.G.S. Uniform with, and forming a Companion Volume to, the same Author's " Metalliferous Minerals and Mining." With 76 Wood Engravings. Second Edition. Crown 8vo, 12s. 6d. cloth.

" It is essentially a practical work, intended primarily for the use of practical men. . . . We do not remember to have met with any English work on mining matters that contains the same amount of information packed in equally convenient form."—*Academy.*

" The book is clearly the result of many years' careful work and thought, and we should be inclined to rank it as among the very best of the handy technical and trades manuals which have recently appeared."—*British Quarterly Review.*

" The volume contains a great mass of practical information carefully methodised and presented in a very intelligible shape."—*Scotsman.*

" The subject matter of the volume will be found of high value by all—and they are a numerous class—who trade in earthy minerals."—*Athenæum.*

Underground Pumping Machinery.

MINE DRAINAGE. Being a Complete and Practical Treatise on Direct-Acting Underground Steam Pumping Machinery, with a Description of a large number of the best known Engines, their General Utility and the Special Sphere of their Action, the Mode of their Application, and their merits compared with other forms of Pumping Machinery. By STEPHEN MICHELL. 8vo, 15s. cloth.

"Will be highly esteemed by colliery owners and lessees, mining engineers, and students generally who require to be acquainted with the best means of securing the drainage of mines. It is a most valuable work, and stands almost alone in the literature of steam pumping machinery."—*Colliery Guardian.*

"Much valuable information is given, so that the book is thoroughly worthy of an extensive circulation amongst practical men and purchasers of machinery."—*Mining Journal.*

Mining Tools.

A MANUAL OF MINING TOOLS. For the Use of Mine Managers, Agents, Students, &c. By WILLIAM MORGANS, Lecturer on Practical Mining at the Bristol School of Mines. 12mo, 3s. cloth boards.

ATLAS OF ENGRAVINGS to Illustrate the above, containing 235 Illustrations of Mining Tools, drawn to scale. 4to, 4s. 6d. cloth.

"Students in the science of mining, and overmen, captains, managers, and viewers may gain practical knowledge and useful hints by the study of Mr. Morgans' manual."—*Colliery Guardian.*

"A valuable work, which will tend materially to improve our mining literature."—*Mining Journal.*

Coal Mining.

COAL AND COAL MINING: A *Rudimentary Treatise on.* By Sir WARINGTON W. SMYTH, M.A., F.R.S., &c., Chief Inspector of the Mines of the Crown. New Edition, Revised and Corrected. With numerous Illustrations. 12mo, 4s. cloth boards.

"As an outline is given of every known coal-field in this and other countries, as well as of the principal methods of working, the book will doubtless interest a very large number of readers."—*Mining Journal.*

Subterraneous Surveying.

SUBTERRANEOUS SURVEYING, *Elementary and Practical Treatise on;* with and without the Magnetic Needle. By THOMAS FENWICK, Surveyor of Mines, and THOMAS BAKER, C.E. Illustrated. 12mo, 3s. cloth boards.

NATURAL AND APPLIED SCIENCE.

Text Book of Electricity.

THE STUDENT'S TEXT-BOOK OF ELECTRICITY. By HENRY M. NOAD, Ph.D., F.R.S., F.C.S. New Edition, carefully Revised. With an Introduction and Additional Chapters, by W. H. PREECE, M.I.C.E., Vice-President of the Society of Telegraph Engineers, &c. With 470 Illustrations. Crown 8vo, 12s. 6d. cloth.

"The original plan of this book has been carefully adhered to so as to make it a reflex of the existing state of electrical science, adapted for students. . . . Discovery seems to have progressed with marvellous strides; nevertheless it has now apparently ceased, and practical applications have commenced their career; and it is to give a faithful account of these that this fresh edition of Dr. Noad's valuable text-book is launched forth."—*Extract from Introduction by W. H. Preece, Esq.*

"We can recommend Dr. Noad's book for clear style, great range of subject, a good index, and a plethora of woodcuts. Such collections as the present are indispensable."—*Athenæum.*

"Dr. Noad's text-book has earned for itself the reputation of a truly scientific manual for the student of electricity, and we gladly hail this new amended edition, which brings it once more to the front. Mr. Preece as reviser, with the assistance of Mr. H. R. Kempe and Mr. J. P. Edwards, has added all the practical results of recent invention and research to the admirable theoretical expositions of the author, so that the book is about as complete and advanced as it is possible for any book to be within the limits of a text-book."—*Telegraphic Journal.*

Electricity.

A MANUAL OF ELECTRICITY: *Including Galvanism, Magnetism, Dia-Magnetism, Electro-Dynamics, Magno-Electricity, and the Electric Telegraph.* By HENRY M. NOAD, Ph.D., F.R.S., F.C.S. Fourth Edition. With 500 Woodcuts. 8vo, £1 4s. cloth.

"The accounts given of electricity and galvanism are not only complete in a scientific sense but, which is a rarer thing, are popular and interesting."—*Lancet.*

"It is worthy of a place in the library of every public institution."—*Mining Journal.*

Electric Light.

ELECTRIC LIGHT : Its Production and Use. Embodying Plain Directions for the Treatment of Voltaic Batteries, Electric Lamps, and Dynamo-Electric Machines. By J. W. URQUHART, C.E., Author of "Electro-plating." Second Edition, with large Additions and 128 Illusts. 7s. 6d. cloth.

"The book is by far the best that we have yet met with on the subject."—*Athenæum*.
"It is the only work at present available which gives, in language intelligible for the most part to the ordinary reader, a general but concise history of the means which have been adopted up to the present time in producing the electric light."—*Metropolitan*.
"The book contains a general account of the means adopted in producing the electric light, not only as obtained from voltaic or galvanic batteries, but treats at length of the dynamo-electric machine in several of its forms."—*Colliery Guardian*.

Electric Lighting.

THE ELEMENTARY PRINCIPLES OF ELECTRIC LIGHT-ING. By ALAN A. CAMPBELL SWINTON, Associate S.T.E. Crown 8vo, 1s. 6d. cloth.

"Anyone who desires a short and thoroughly clear exposition of the elementary principles of electric-lighting cannot do better than read this little work."—*Bradford Observer*.

Dr. Lardner's School Handbooks.

NATURAL PHILOSOPHY FOR SCHOOLS. By Dr. LARDNER. 328 Illustrations. Sixth Edition. One Vol., 3s. 6d. cloth.

"A very convenient class-book for junior students in private schools. It is intended to convey, in clear and precise terms, general notions of all the principal divisions of Physical Science."—*British Quarterly Review*.

ANIMAL PHYSIOLOGY FOR SCHOOLS. By Dr. LARDNER. With 190 Illustrations. Second Edition. One Vol., 3s. 6d. cloth.

"Clearly written, well arranged, and excellently illustrated."—*Gardener's Chronicle*.

Dr. Lardner's Electric Telegraph.

THE ELECTRIC TELEGRAPH. By Dr. LARDNER. Revised and Re-written by E. B. BRIGHT, F.R.A.S. 140 Illustrations. Small 8vo, 2s. 6d. cloth.

"One of the most readable books extant on the Electric Telegraph."—*English Mechanic*.

Storms.

STORMS: Their Nature, Classification, and Laws; with the Means of Predicting them by their Embodiments, the Clouds. By WM. BLASIUS. With Coloured Plates and Woodcuts. Crown 8vo, 10s. 6d. cloth.

"A very readable book. . . . The fresh facts contained in its pages, collected with evident care, form a useful repository for meteorologists in the study of atmospherical disturbances. . . . The book will pay perusal as being the production of one who gives evidence of acute observation."—*Nature*.

The Blowpipe.

THE BLOWPIPE IN CHEMISTRY, MINERALOGY, AND GEOLOGY. Containing all known Methods of Anhydrous Analysis, many Working Examples, and Instructions for Making Apparatus. By Lieut.-Colonel W. A. Ross, R.A. With 120 Illustrations. Cr. 8vo, 5s. 6d. cloth.

"The student who goes conscientiously through the course of experimentation here laid down will gain a better insight into inorganic chemistry and mineralogy than if he had 'got up' any of the best text-books of the day, and passed any number of examinations in their contents."—*Chemical News*.

The Military Sciences.

AIDE-MEMOIRE TO THE MILITARY SCIENCES. Framed from Contributions of Officers and others connected with the different Services. Originally edited by a Committee of the Corps of Royal Engineers. Second Edition, most carefully revised by an Officer of the Corps, with many Additions; containing nearly 350 Engravings and many hundred Woodcuts. Three Vols., royal 8vo, extra cloth boards, and lettered, £4 10s.

"A compendious encyclopædia of military knowledge, to which we are greatly indebted."—*Edinburgh Review*.

Field Fortification.

A TREATISE ON FIELD FORTIFICATION, THE ATTACK OF FORTRESSES, MILITARY MINING, AND RECONNOITRING. By Colonel I. S. MACAULAY, late Professor of Fortification in the R.M.A., Woolwich. Sixth Edition, crown 8vo, cloth, with separate Atlas of 12 Plates, 12s.

Temperaments.

OUR TEMPERAMENTS, THEIR STUDY AND THEIR TEACHING. A Popular Outline. By ALEXANDER STEWART, F.R.C.S. Edin. In one large 8vo volume, with 30 Illustrations, including A Selection from Lodge's "Historical Portraits," showing the Chief Forms of Faces. Price 15s. cloth, gilt top.

"The book is exceedingly interesting, even for those who are not systematic students of anthropology.... To those who think the proper study of mankind is man, it will be full of attraction."—*Daily Telegraph.*

"The author's object is to enable a student to read a man's temperament in his aspect. The work is well adapted to its end. It is worthy of the attention of students of human nature."—*Scotsman.*

"The volume is heavy to hold, but light to read. Though the author has treated his subject exhaustively, he writes in a popular and pleasant manner that renders it attractive to the general reader."—*Punch.*

Pneumatics and Acoustics.

PNEUMATICS: *including Acoustics and the Phenomena of Wind Currents,* for the Use of Beginners. By CHARLES TOMLINSON, F.R.S., F.C.S., &c. Fourth Edition, Enlarged. With numerous Illustrations. 12mo, 1s. 6d. cloth. [*Just published.*

"Beginners in the study of this important application of science could not have a better manual."—*Scotsman.*

"A valuable and suitable text-book for students of Acoustics and the Phenomena of Wind Currents."—*Schoolmaster.*

Conchology.

A MANUAL OF THE MOLLUSCA: *Being a Treatise on Recent and Fossil Shells.* By S. P. WOODWARD, A.L.S., F.G.S., late Assistant Palæontologist in the British Museum. Fifth Edition. With an Appendix on *Recent and Fossil Conchological Discoveries,* by RALPH TATE A.L.S., F.G.S. Illustrated by A. N. WATERHOUSE and JOSEPH WILSON LOWRY. With 23 Plates and upwards of 300 Woodcuts. Crown 8vo, 7s. 6d. cloth boards.

"A most valuable storehouse of conchological and geological information."—*Science Gossip.*

Astronomy.

ASTRONOMY. By the late Rev. ROBERT MAIN, M.A., F.R.S., formerly Radcliffe Observer at Oxford. Third Edition, Revised and Corrected to the present time, by WILLIAM THYNNE LYNN, B.A., F.R.A.S., formerly of the Royal Observatory, Greenwich. 12mo, 2s. cloth limp.

"A sound and simple treatise, very carefully edited, and a capital book for beginners."—*Knowledge.*

"Accurately brought down to the requirements of the present time by Mr. Lynn."—*Educational Times.*

Geology.

RUDIMENTARY TREATISE ON GEOLOGY, PHYSICAL AND HISTORICAL. Consisting of "Physical Geology," which sets forth the leading Principles of the Science; and "Historical Geology," which treats of the Mineral and Organic Conditions of the Earth at each successive epoch, especial reference being made to the British Series of Rocks. By RALPH TATE, A.L.S., F.G.S., &c., &c. With 250 Illustrations. 12mo, 5s. cloth boards.

"The fulness of the matter has elevated the book into a manual. Its information is exhaustive and well arranged."—*School Board Chronicle.*

Geology and Genesis.

THE TWIN RECORDS OF CREATION; *or, Geology and Genesis: their Perfect Harmony and Wonderful Concord.* By GEORGE W. VICTOR LE VAUX. Numerous Illustrations. Fcap. 8vo, 5s. cloth.

"A valuable contribution to the evidences of Revelation, and disposes very conclusively of the arguments of those who would set God's Works against God's Word. No real difficulty is shirked, and no sophistry is left unexposed."—*The Rock.*

"The remarkable peculiarity of this author is that he combines an unbounded admiration of science with an unbounded admiration of the Written record. The two impulses are balanced to a nicety; and the consequence is that difficulties, which to minds less evenly poised would be serious, find immediate solutions of the happiest kinds."—*London Review.*

DR. LARDNER'S HANDBOOKS OF NATURAL PHILOSOPHY.

THE HANDBOOK OF MECHANICS. Enlarged and almost rewritten by BENJAMIN LOEWY, F.R.A.S. With 378 Illustrations. Post 8vo, 6s. cloth.

"The perspicuity of the original has been retained, and chapters which had become obsolete have been replaced by others of more modern character. The explanations throughout are studiously popular, and care has been taken to show the application of the various branches of physics to the industrial arts, and to the practical business of life."—*Mining Journal.*

"Mr. Loewy has carefully revised the book, and brought it up to modern requirements."—*Nature.*

"Natural philosophy has had few exponents more able or better skilled in the art of popularising the subject than Dr. Lardner; and Mr. Loewy is doing good service in fitting this treatise, and the others of the series, for use at the present time."—*Scotsman.*

THE HANDBOOK OF HYDROSTATICS AND PNEUMATICS. New Edition, Revised and Enlarged, by BENJAMIN LOEWY, F.R.A.S. With 236 Illustrations. Post 8vo, 5s. cloth.

"For those 'who desire to attain an accurate knowledge of physical science without the profound methods of mathematical investigation,' this work is not merely intended, but well adapted."—*Chemical News.*

"The volume before us has been carefully edited, augmented to nearly twice the bulk of the former edition, and all the most recent matter has been added. . . . It is a valuable text-book."—*Nature.*

"Candidates for pass examinations will find it, we think, specially suited to their requirements."—*English Mechanic.*

THE HANDBOOK OF HEAT. Edited and almost entirely rewritten by BENJAMIN LOEWY, F.R.A.S., &c. 117 Illustrations. Post 8vo, 6s. cloth.

"The style is always clear and precise, and conveys instruction without leaving any cloudiness or lurking doubts behind."—*Engineering.*

"A most exhaustive book on the subject on which it treats, and is so arranged that it can be understood by all who desire to attain an accurate knowledge of physical science. . . . . Mr. Loewy has included all the latest discoveries in the varied laws and effects of heat."—*Standard.*

"A complete and handy text-book for the use of students and general readers."—*English Mechanic.*

THE HANDBOOK OF OPTICS. By DIONYSIUS LARDNER, D.C.L., formerly Professor of Natural Philosophy and Astronomy in University College, London. New Edition. Edited by T. OLVER HARDING, B.A. Lond., of University College, London. With 298 Illustrations. Small 8vo, 448 pages, 5s. cloth.

"Written by one of the ablest English scientific writers, beautifully and elaborately illustrated."—*Mechanic's Magazine.*

THE HANDBOOK OF ELECTRICITY, MAGNETISM, AND ACOUSTICS. By Dr. LARDNER. Ninth Thousand. Edit. by GEORGE CAREY FOSTER, B.A., F.C.S. With 400 Illustrations. Small 8vo, 5s. cloth.

"The book could not have been entrusted to anyone better calculated to preserve the terse and lucid style of Lardner, while correcting his errors and bringing up his work to the present state of scientific knowledge."—*Popular Science Review.*

*** *The above Five Volumes, though each is Complete in itself, form* A COMPLETE COURSE OF NATURAL PHILOSOPHY.

Dr. Lardner's Handbook of Astronomy.

THE HANDBOOK OF ASTRONOMY. Forming a Companion to the "Handbook of Natural Philosophy." By DIONYSIUS LARDNER, D.C.L., formerly Professor of Natural Philosophy and Astronomy in University College, London. Fourth Edition. Revised and Edited by EDWIN DUNKIN, F.R.A.S., Royal Observatory, Greenwich. With 38 Plates and upwards of 100 Woodcuts. In One Vol., small 8vo, 550 pages, 9s. 6d. cloth.

"Probably no other book contains the same amount of information in so compendious and well-arranged a form—certainly none at the price at which this is offered to the public."—*Athenæum.*

"We can do no other than pronounce this work a most valuable manual of astronomy, and we strongly recommend it to all who wish to acquire a general—but at the same time correct—acquaintance with this sublime science."—*Quarterly Journal of Science.*

"One of the most deservedly popular books on the subject . . . We would recommend not only the student of the elementary principles of the science, but he who aims at mastering the higher and mathematical branches of astronomy, not to be without this work beside him."—*Practical Magazine.*

DR. LARDNER'S MUSEUM OF SCIENCE AND ART.

THE MUSEUM OF SCIENCE AND ART. Edited by DIONYSIUS LARDNER, D.C.L., formerly Professor of Natural Philosophy and Astronomy in University College, London. With upwards of 1,200 Engravings on Wood. In 6 Double Volumes, £1 1s., in a new and elegant cloth binding; or handsomely bound in half-morocco, 31s. 6d.

Contents:

The Planets: Are they Inhabited Worlds?—Weather Prognostics — Popular Fallacies in Questions of Physical Science—Latitudes and Longitudes — Lunar Influences — Meteoric Stones and Shooting Stars—Railway Accidents—Light—Common Things: Air—Locomotion in the United States—Cometary Influences—Common Things: Water—The Potter's Art—Common Things: Fire — Locomotion and Transport, their Influence and Progress—The Moon — Common Things: The Earth—The Electric Telegraph — Terrestrial Heat — The Sun—Earthquakes and Volcanoes—Barometer, Safety Lamp, and Whitworth's Micrometric Apparatus—Steam—The Steam Engine—The Eye—The Atmosphere — Time — Common Things: Pumps—Common Things: Spectacles, the Kaleidoscope — Clocks and Watches — Microscopic Drawing and Engraving—Locomotive — Thermometer — New Planets: Leverrier and Adams's Planet—Magnitude and Minuteness—Common Things: The Almanack—Optical Images—How to observe the Heavens — Common Things: The Looking-glass — Stellar Universe—The Tides—Colour—Common Things: Man—Magnifying Glasses—Instinct and Intelligence—The Solar Microscope—The Camera Lucida—The Magic Lantern—The Camera Obscura—The Microscope—The White Ants: Their Manners and Habits—The Surface of the Earth, or First Notions of Geography—Science and Poetry—The Bee—Steam Navigation — Electro-Motive Power — Thunder, Lightning, and the Aurora Borealis—The Printing Press—The Crust of the Earth—Comets—The Stereoscope—The Pre-Adamite Earth—Eclipses—Sound.

*** OPINIONS OF THE PRESS.

"This series, besides affording popular but sound instruction on scientific subjects, with which the humblest man in the country ought to be acquainted, also undertakes that teaching of 'Common Things' which every well-wisher of his kind is anxious to promote. Many thousand copies of this serviceable publication have been printed, in the belief and hope that the desire for instruction and improvement widely prevails; and we have no fear that such enlightened faith will meet with disappointment."—*Times.*

"A cheap and interesting publication, alike informing and attractive. The papers combine subjects of importance and great scientific knowledge, considerable inductive powers, and a popular style of treatment."—*Spectator.*

"The 'Museum of Science and Art' is the most valuable contribution that has ever been made to the Scientific Instruction of every class of society."—SIR DAVID BREWSTER, in the *North British Review.*

"Whether we consider the liberality and beauty of the Illustrations, the charm of the writing, or the durable interest of the matter, we must express our belief that there is hardly to be found among the new books one that would be welcomed by people of so many ages and classes as a valuable present."—*Examiner.*

*** *Separate books formed from the above, suitable for Workmen's Libraries, Science Classes, etc.*

Common Things Explained. Containing Air, Earth, Fire, Water, Time, Man, the Eye, Locomotion, Colour, Clocks and Watches, &c. 233 Illustrations, cloth gilt, 5s.

The Microscope. Containing Optical Images, Magnifying Glasses, Origin and Description of the Microscope, Microscopic Objects, the Solar Microscope, Microscopic Drawing and Engraving, &c. 147 Illustrations, cloth gilt, 2s.

Popular Geology. Containing Earthquakes and Volcanoes, the Crust of the Earth, &c. 201 Illustrations, cloth gilt, 2s. 6d.

Popular Physics. Containing Magnitude and Minuteness, the Atmosphere, Meteoric Stones, Popular Fallacies, Weather Prognostics, the Thermometer, the Barometer, Sound, &c. 85 Illustrations, cloth gilt, 2s. 6d.

Steam and its Uses. Including the Steam Engine, the Locomotive, and Steam Navigation. 89 Illustrations, cloth gilt, 2s.

Popular Astronomy. Containing How to observe the Heavens—The Earth, Sun, Moon, Planets, Light, Comets, Eclipses, Astronomical Influences, &c. 182 Illustrations, 4s. 6d.

The Bee and White Ants: Their Manners and Habits. With Illustrations of Animal Instinct and Intelligence. 135 Illustrations, cloth gilt, 2s.

The Electric Telegraph Popularized. To render intelligible to all who can Read, irrespective of any previous Scientific Acquirements, the various forms of Telegraphy in Actual Operation. 100 Illustrations, cloth gilt, 1s. 6d.

COUNTING-HOUSE WORK, TABLES, etc.

Accounts for Manufacturers.

FACTORY ACCOUNTS: Their Principles and Practice. A Handbook for Accountants and Manufacturers, with Appendices on the Nomenclature of Machine Details; the Income Tax Acts; the Rating of Factories; Fire and Boiler Insurance; the Factory and Workshop Acts, &c., including also a Glossary of Terms and a large number of Specimen Rulings. By EMILE GARCKE and J. M. FELLS. Second Edition. Demy 8vo, 250 pages, price 10s. 6d. strongly bound. [*Just published.*

"One of the most important works ever published dealing with these matters. The authors have treated the subject from the standpoint of the factory, as practical men speaking to practical men, and not, as has been too often the case, as schoolmasters to schoolboys."—*Electrician.*

"A very interesting description of the requirements of Factory Accounts. . . . the principle of assimilating the Factory Accounts to the general commercial books is one which we thoroughly agree with."—*Accountants' Journal.*

"Characterised by extreme thoroughness. There are few owners of Factories who would not derive great benefit from the perusal of this most admirable work."—*Local Government Chronicle.*

Foreign Commercial Correspondence.

THE FOREIGN COMMERCIAL CORRESPONDENT: Being Aids to Commercial Correspondence in Five Languages—English, French, German, Italian and Spanish. By CHARLES E. BAKER. Crown 8vo, price about 5s. [*In preparation.*

Intuitive Calculations.

THE COMPENDIOUS CALCULATOR; or, Easy and Concise Methods of Performing the various Arithmetical Operations required in Commercial and Business Transactions, together with Useful Tables. By DANIEL O'GORMAN. Corrected and Extended by J. R. YOUNG, formerly Professor of Mathematics at Belfast College. Twenty-sixth Edition, carefully Revised by C. NORRIS. Fcap. 8vo, 3s. 6d. strongly half-bound in leather.

"It would be difficult to exaggerate the usefulness of a book like this to everyone engaged in commerce or manufacturing industry. It is crammed full of rules and formulæ for shortening and employing calculations."—*Knowledge.*

"Supplies special and rapid methods for all kinds of calculations. Of great utility to persons engaged in any kind of commercial transactions."—*Scotsman.*

Modern Metrical Units and Systems.

MODERN METROLOGY: A Manual of the Metrical Units and Systems of the Present Century. With an Appendix containing a proposed English System. By LOWIS D'A. JACKSON, A.M. Inst. C.E., Author of "Aid to Survey Practice," &c. Large crown 8vo, 12s. 6d. cloth.

"The author has brought together much valuable and interesting information. . . . We cannot but recommend the work to the consideration of all interested in the practical reform of our weights and measures."—*Nature.*

"For exhaustive tables of equivalent weights and measures of all sorts, and for clear demonstrations of the effects of the various systems that have been proposed or adopted, Mr. Jackson's treatise is without a rival."—*Academy.*

The Metric System and the British Standards.

A SERIES OF METRIC TABLES, in which the British Standard Measures and Weights are compared with those of the Metric System at present in Use on the Continent. By C. H. DOWLING, C.E. 8vo, 10s. 6d. strongly bound.

"Their accuracy has been certified by Professor Airy, the Astronomer-Royal."—*Builder.*

"Mr. Dowling's Tables are well put together as a ready-reckoner for the conversion of one system into the other."—*Athenæum.*

Iron and Metal Trades' Calculator.

THE IRON AND METAL TRADES' COMPANION. For expeditiously ascertaining the Value of any Goods bought or sold by Weight, from 1s. per cwt. to 112s. per cwt., and from one farthing per pound to one shilling per pound. Each Table extends from one pound to 100 tons. To which are appended Rules on Decimals, Square and Cube Root, Mensuration of Superficies and Solids, &c.; Tables of Weights of Materials, and other Useful Memoranda. By THOS. DOWNIE. 396 pp., 9s. Strongly bound in leather.

"A most useful set of tables, and will supply a want, for nothing like them ever existed."—*Building News.*

"Although specially adapted to the iron and metal trades, the tables will be found useful in every other business in which merchandise is bought and sold by weight."—*Railway News.*

Calculator for Numbers and Weights Combined.

THE COMBINED NUMBER AND WEIGHT CALCULATOR. Containing upwards of 250,000 Separate Calculations, showing at a glance the value at 421 different rates, ranging from $\frac{1}{16}$th of a Penny to 20s. each, or per cwt., and £20 per ton, of any number of articles consecutively, from 1 to 470.—Any number of cwts., qrs., and lbs., from 1 cwt. to 470 cwts.—Any number of tons, cwts., qrs., and lbs., from 1 to 23½ tons. By WILLIAM CHADWICK, Public Accountant. Imp. 8vo, 30s. strongly bound for Office wear and tear.

☞ *This comprehensive and entirely unique and original Calculator is adapted for the use of Accountants and Auditors, Railway Companies, Canal Companies, Shippers, Shipping Agents, General Carriers, etc. Ironfounders, Brassfounders, Metal Merchants, Iron Manufacturers, Ironmongers, Engineers, Machinists, Boiler Makers, Millwrights, Roofing, Bridge and Girder Makers, Colliery Proprietors, etc. Timber Merchants, Builders, Contractors, Architects, Surveyors, Auctioneers, Valuers, Brokers, Mill Owners and Manufacturers, Mill Furnishers, Merchants and General Wholesale Tradesmen.*

*** OPINIONS OF THE PRESS.

"The book contains the answers to questions, and not simply a set of ingenious puzzle methods of arriving at results. It is as easy of reference for any answer or any number of answers as a dictionary, and the references are even more quickly made. For making up accounts or estimates, the book must prove invaluable to all who have any considerable quantity of calculations involving price and measure in any combination to do."—*Engineer.*

"The most complete and practical ready reckoner which it has been our fortune yet to see. It is difficult to imagine a trade or occupation in which it could not be of the greatest use, either in saving human labour or in checking work."—*The Miller.*

"The most perfect work of the kind yet prepared."—*Glasgow Herald.*

Comprehensive Weight Calculator.

THE WEIGHT CALCULATOR. Being a Series of Tables upon a New and Comprehensive Plan, exhibiting at One Reference the exact Value of any Weight from 1 lb. to 15 tons, at 300 Progressive Rates, from 1d. to 168s. per cwt., and containing 186,000 Direct Answers, which, with their Combinations, consisting of a single addition (mostly to be performed at sight), will afford an aggregate of 10,266,000 Answers; the whole being calculated and designed to ensure correctness and promote despatch. By HENRY HARBEN, Accountant. Fourth Edition, carefully Corrected. Royal 8vo, strongly half-bound, £1 5s. [*Just published.*

"A practical and useful work of reference for men of business generally; it is the best of the kind we have seen."—*Ironmonger.*

"Of priceless value to business men. Its accuracy and completeness have secured for it a reputation which renders it quite unnecessary for us to say one word in its praise. It is a necessary book in all mercantile offices."—*Sheffield Independent.*

Comprehensive Discount Guide.

THE DISCOUNT GUIDE. Comprising several Series of Tables for the use of Merchants, Manufacturers, Ironmongers, and others, by which may be ascertained the exact Profit arising from any mode of using Discounts, either in the Purchase or Sale of Goods, and the method of either Altering a Rate of Discount or Advancing a Price, so as to produce, by one operation, a sum that will realise any required profit after allowing one or more Discounts: to which are added Tables of Profit or Advance from 1¼ to 90 per cent., Tables of Discount from 1¼ to 98¾ per cent., and Tables of Commission, &c., from ⅛ to 10 per cent. By HENRY HARBEN, Accountant, Author of "The Weight Calculator." New Edition, carefully Revised and Corrected. Demy 8vo, 544 pp. half-bound, £1 5s.

"A book such as this can only be appreciated by business men, to whom the saving of time means saving of money. We have the high authority of Professor J. R. Young that the tables throughout the work are constructed upon strictly accurate principles. The work must prove of great value to merchants, manufacturers, and general traders."—*British Trade Journal.*

Iron Shipbuilders' and Merchants' Weight Tables.

IRON-PLATE WEIGHT TABLES: For Iron Shipbuilders, Engineers and Iron Merchants. Containing the Calculated Weights of upwards of 150,000 different sizes of Iron Plates, from 1 foot by 6 in. by ¼ in. to 10 feet by 5 feet by 1 in. Worked out on the basis of 40 lbs. to the square foot of Iron of 1 inch in thickness. Carefully compiled and thoroughly Revised by H. BURLINSON and W. H. SIMPSON. Oblong 4to, 25s. half-bound.

"This work will be found of great utility. The authors have had much practical experience of what is wanting in making estimates; and the use of the book will save much time in making elaborate calculations.—*English Mechanic.*

INDUSTRIAL AND USEFUL ARTS.

Soap-making.

THE ART OF SOAP-MAKING: A Practical Handbook of the Manufacture of Hard and Soft Soaps, Toilet Soaps, etc. Including many New Processes, and a Chapter on the Recovery of Glycerine from Waste Leys. By ALEXANDER WATT, Author of "Electro-Metallurgy Practically Treated," &c. With numerous Illustrations. Third Edition, Revised. Crown 8vo, 7s. 6d. cloth.

"The work will prove very useful, not merely to the technological student, but to the practical soap-boiler who wishes to understand the theory of his art."—*Chemical News.*

"Really an excellent example of a technical manual, entering, as it does, thoroughly and exhaustively both into the theory and practice of soap manufacture. The book is well and honestly done, and deserves the considerable circulation with which it will doubtless meet."—*Knowledge.*

"Mr. Watt's book is a thoroughly practical treatise on an art which has almost no literature in our language. We congratulate the author on the success of his endeavour to fill a void in English technical literature."—*Nature.*

Leather Manufacture.

THE ART OF LEATHER MANUFACTURE. Being a Practical Handbook, in which the Operations of Tanning, Currying, and Leather Dressing are fully Described, and the Principles of Tanning Explained, and many Recent Processes introduced; as also Methods for the Estimation of Tannin, and a Description of the Arts of Glue Boiling, Gut Dressing, &c. By ALEXANDER WATT, Author of "Soap-Making," "Electro-Metallurgy," &c. With numerous Illustrations. Second Edition. Crown 8vo, 9s. cloth. [*Just published.*

"A sound, comprehensive treatise on tanning and its accessories. The book is an eminently valuable production, which redounds to the credit of both author and publishers."—*Chemical Review.*

"This volume is technical without being tedious, comprehensive and complete without being prosy, and it bears on every page the impress of a master hand. We have never come across a better trade treatise, nor one that so thoroughly supplied an absolute want."—*Shoe and Leather Trades' Chronicle.*

Boot and Shoe Making.

THE ART OF BOOT AND SHOE-MAKING. A Practical Handbook, including Measurement, Last-Fitting, Cutting-Out, Closing and Making, with a Description of the most approved Machinery employed. By JOHN B. LENO, late Editor of *St. Crispin*, and *The Boot and Shoe-Maker*. With numerous Illustrations. Second Edition. Crown 8vo, 2s. 6d. cloth. [*Just published.*

"This excellent treatise is by far the best work ever written on the subject. A new wo k, embracing all modern improvements, was much wanted. This want is now satisfied. The chapter on clicking, which shows how waste may be prevented, will save fifty times the price of the book"—*Scottish Leather Trader.*

"This volume is replete with matter well worthy the perusal of boot and shoe manufacturers and experienced craftsmen, and instructive and valuable in the highest degree to all young beginners and craftsmen in the trade of which it treats."—*Leather Trades' Circular.*

Dentistry.

MECHANICAL DENTISTRY: A Practical Treatise on the Construction of the various kinds of Artificial Dentures. Comprising also Useful Formulæ, Tables and Receipts for Gold Plate, Clasps, Solders, &c. &c. By CHARLES HUNTER. Third Edition, Revised. With upwards of 100 Wood Engravings. Crown 8vo, 3s. 6d. cloth. [*Just published.*

"The work is very practical."—*Monthly Review of Dental Surgery.*

"We can strongly recommend Mr. Hunter's treatise to all students preparing for the profession of dentistry, as well as to every mechanical dentist."—*Dublin Journal of Medical Science.*

Wood Engraving.

A PRACTICAL MANUAL OF WOOD ENGRAVING. With a Brief Account of the History of the Art. By WILLIAM NORMAN BROWN. With numerous Illustrations. Crown 8vo, 2s. cloth.

"The author deals with the subject in a thoroughly practical and easy series of representative lessons."—*Paper and Printing Trades Journal.*

"The book is clear and complete, and will be useful to anyone wanting to understand the first elements of the beautiful art of wood engraving."—*Graphic.*

Paper Making.

A TREATISE ON PAPER; with an Outline of its Manufacture, Complete Tables of Sizes, etc. For Printers and Stationers. By RICHARDSON PARKINSON. 8vo, 3s. cloth; 2s. 6d. paper wrapper.

"An admirable handbook by a man who understands his subject."—*Printers' Register.*

LOCKWOOD'S HANDYBOOKS FOR HANDICRAFTS.

☞ *These Handybooks are written to supply Handicraftsmen with information on workshop practice, and are intended to convey, in plain language, technical knowledge of the several crafts. Workshop terms are used, and workshop practice described, the text being freely illustrated with drawings of modern tools, appliances and processes, useful alike to the young beginner and to the old hand, whose range of experience has been narrowed under a system of divided labour, as well as to amateurs.*

☞ *The following Volumes are already published.*

Metal Turning.

THE METAL TURNER'S HANDYBOOK. A Practical Manual for Workers at the Foot-Lathe: Embracing Information on the Tools, Appliances and Processes employed in Metal Turning. By PAUL N. HASLUCK, A.I.M.E., Author of "Lathe-Work." With upwards of One Hundred Illustrations. Second Edition, Revised. Cr. 8vo, 2s. cloth. [*Just published.*

"Altogether admirably adapted to initiate students into the art of turning."—*Leicester Post.*
"Clearly and concisely written, excellent in every way, we heartily commend it to all interested in metal turning."—*Mechanical World.*
"With the assistance of a clever master, a clear and vivid expounder, and an abundance of illustrations, the work lets handicraftsmen know what are the resources of the turning-lathe and how these may be developed."—*Dundee Advertiser.*

Wood Turning.

THE WOOD TURNER'S HANDYBOOK. A Practical Manual for Workers at the Lathe: Embracing Information on the Tools, Appliances and Processes Employed in Wood Turning. By PAUL N. HASLUCK, A.I.M.E, Author of "Lathe-Work," "The Metal Turner's Handybook," &c. With upwards of One Hundred Illustrations. Crown 8vo, 2s. cloth. [*Just published.*

"The volume is well and clearly written in a lucid style, and all the instructions are fully given. It will be found of great value to workmen and amateurs, and forms a safe and reliable guide to every branch of the lathe manipulation."—*Carpenter and Builder.*
"An excellent manual for workers at the lathe."—*Glasgow Herald.*
"We recommend the book to young turners and amateurs. A multitude of workmen have hitherto sought in vain for a manual of this special industry."—*Mechanical World.*

Watch Repairing.

THE WATCH JOBBER'S HANDYBOOK. A Practical Manual on Cleaning, Repairing and Adjusting. Embracing Information on the Tools, Materials, Appliances and Processes Employed in Watchwork. By PAUL N. HASLUCK, A.I.M.E., Author of "Lathe-Work," "The Metal Turner's Handy-Book," "The Wood Turner's Handybook," &c. With upwards of One Hundred Illustrations. Crown 8vo, 2s. cloth. [*Just published.*

"Written in a clear style exactly suited to beginners and amateurs. We heartily recommend it."—*Practical Engineer.*
"We recommend it to craftsmen in watchmaking as a useful and well-written grammar of their art."—*Scotsman.*
"All young persons connected with the trade should acquire and study this excellent, and at the same time, inexpensive work."—*Clerkenwell Chronicle.*

Pattern Making.

THE PATTERN MAKER'S HANDYBOOK. A Practical Manual, embracing Information on the Tools, Materials and Appliances employed in Constructing Patterns for Founders. By PAUL N. HASLUCK, A.I.M.E. With One Hundred Illustrations. Cr. 8vo, 2s. cloth. [*Just published.*

"Mr. Hasluck's 'Lathe Work' and kindred productions have acquired a high reputation. His new volume, 'Pattern Making,' contains invaluable advice, and furnishes the studious workman with a very large amount of practical information."—*Lloyd's News.*
"Especially useful to the beginner. We commend it to all who are interested in the counsels it so ably gives."—*Colliery Guardian.*
"This handy volume contains sound information of considerable value to students and artificers."—*Hardware Trade Journal.*

Mechanical Manipulation.

THE MECHANIC'S WORKSHOP HANDYBOOK. A Practical Manual on Mechanical Manipulation. Embracing Information on various Handicraft Processes, with Useful Notes and Miscellaneous Memoranda. By PAUL N. HASLUCK, A.I.M.E. Author of "Lathe-Work," "The Metal Turner's Handybook," "The Wood Turner's Handybook," &c. Crown 8vo, 2s. cloth. [*Just ready*

INDUSTRIAL AND USEFUL ARTS. 33

Electrolysis of Gold, Silver, Copper, etc.

ELECTRO-DEPOSITION : *A Practical Treatise on the Electrolysis of Gold, Silver, Copper, Nickel, and other Metals and Alloys.* With descriptions of Voltaic Batteries, Magnet and Dynamo-Electric Machines, Thermopiles, and of the Materials and Processes used in every Department of the Art, and several Chapters on ELECTRO-METALLURGY. By ALEXANDER WATT, Author of "Electro-Metallurgy," &c. With numerous Illustrations. Second Edition, Revised and Corrected. Crown 8vo, 9s. cloth.
[*Just published.*

"Evidently written by a practical man who has spent a long period of time in electro-plate workshops. The information given respecting the details of workshop manipulation is remarkably complete. . . . Mr. Watt's book will prove of great value to electro-depositors, jewellers, and various other workers in metal."—*Nature.*

"Eminently a book for the practical worker in electro-deposition. It contains minute and practical descriptions of methods, processes and materials as actually pursued and used in the workshop. Mr. Watt's book recommends itself to all interested in its subjects."—*Engineer.*

Electro-Metallurgy.

ELECTRO-METALLURGY; *Practically Treated.* By ALEXANDER WATT, F.R.S.S.A. Eighth Edition, Revised, with Additional Matter and Illustrations, including the most recent Processes. 12mo, 3s. 6d. cloth boards.

"From this book both amateur and artisan may learn everything necessary for the successful prosecution of electroplating."—*Iron.*

Electroplating.

ELECTROPLATING : *A Practical Handbook.* By J. W. URQUHART, C.E. With numerous Illustrations. Crown 8vo, 5s. cloth.

"The information given appears to be based on direct personal knowledge. . . . Its science is sound and the style is always clear."—*Athenæum.*

Electrotyping.

ELECTROTYPING : *The Reproduction and Multiplication of Printing Surfaces and Works of Art by the Electro-deposition of Metals.* By J. W. URQUHART, C.E. Crown 8vo, 5s. cloth.

"The book is thoroughly practical. The reader is, therefore, conducted through the leading laws of electricity, then through the metals used by electrotypers, the apparatus, and the depositing processes, up to the final preparation of the work."—*Art Journal.*

"We can recommend this treatise, not merely to amateurs, but to those actually engaged in the trade."—*Chemical News.*

Goldsmiths' Work.

THE GOLDSMITH'S HANDBOOK. By GEORGE E. GEE, Jeweller, &c. Third Edition, considerably Enlarged. 12mo, 3s. 6d. cloth boards.

"A good, sound, technical educator, and will be generally accepted as an authority. It exactly fulfils the purpose intended."—*Horological Journal.*

"Will speedily become a standard book which few will care to be without."—*Jeweller and Metalworker.*

Silversmiths' Work.

THE SILVERSMITH'S HANDBOOK. By GEORGE E. GEE, Jeweller, &c. Second Edition, Revised, with numerous Illustrations. 12mo, 3s. 6d. cloth boards.

"The chief merit of the work is its practical character. . . . The workers in the trade will speedily discover its merits when they sit down to study it."—*English Mechanic.*

"This work forms a valuable sequel to the author's 'Goldsmith's Handbook.'"—*Silversmiths' Trade Journal.*

*** *The above two works together, strongly half-bound, price 7s.*

Textile Manufacturers' Tables.

UNIVERSAL TABLES OF TEXTILE STRUCTURE. For the use of Manufacturers in every branch of Textile Trade. By JOSEPH EDMONDSON. Oblong folio, strongly bound in cloth, price 7s. 6d.

☞ *The principle on which the tables are founded is well known, and much used in the muslin manufacture, but the intricacy of the calculations hitherto required (especially where warp and weft differ in counts and in the closeness of the threads) has prevented its general application. By these tables all the adjustments may be made without calculation. Mere references to the proper places bring out the required information.*

"Immense labour has been bestowed on the work by the author. The tables are adapted to every mode of numbering yarns and setts, and apply to all the branches of textile manufacture."—*Textile Recorder.*

Horology.

A TREATISE ON MODERN HOROLOGY, in Theory and Practice. Translated from the French of CLAUDIUS SAUNIER, ex-Director of the School of Horology at Macon, by JULIEN TRIPPLIN, F.R.A.S., Besancon Watch Manufacturer, and EDWARD RIGG, M.A., Assayer in the Royal Mint. With Seventy-eight Woodcuts and Twenty-two Coloured Copper Plates. Second Edition. Super-royal 8vo, £2 2s. cloth; £2 10s. half-calf.

"There is no horological work in the English language at all to be compared to this production of M. Saunier's for clearness and completeness. It is alike good as a guide for the student and as a reference for the experienced horologist and skilled workman."—*Horological Journal.*

"The latest, the most complete, and the most reliable of those literary productions to which continental watchmakers are indebted for the mechanical superiority over their English brethren —in fact, the Book of Books, is M. Saunier's 'Treatise.'"—*Watchmaker, Jeweller and Silversmith.*

"This magnificent treatise is one of the most valuable and comprehensive contributions to the literature of horological art and science ever produced, and cannot be too highly commended. It is a perfect cyclopædia of watch and clockmaking."—*The Coventry Watch and Clockmaker.*

Watchmaking.

THE WATCHMAKER'S HANDBOOK. Intended as a Workshop Companion for those engaged in Watchmaking and the Allied Mechanical Arts. Translated from the French of CLAUDIUS SAUNIER, and considerably Enlarged by JULIEN TRIPPLIN, F.R.A.S., Vice-President of the Horological Institute, and EDWARD RIGG, M.A., Assayer in the Royal Mint. With Numerous Woodcuts and Fourteen Copper Plates. Second Edition, Revised. With Appendix. Crown 8vo, 9s. cloth. [*Just published.*

"Each part is truly a treatise in itself. The arrangement is good and the language is clear and concise. It is an admirable guide for the young watchmaker."—*Engineering.*

"It is impossible to speak too highly of its excellence. It fulfils every requirement in a handbook intended for the use of a workman. Should be found in every workshop."—*Watch and Clockmaker.*

"This book contains an immense number of practical details bearing on the daily occupation of a watchmaker, and it will be found of great use to an army of workers."—*Watchmaker and Metalworker* (Chicago).

CHEMICAL MANUFACTURES & COMMERCE.

The Alkali Trade, Sulphuric Acid, etc.

A MANUAL OF THE ALKALI TRADE, including the Manufacture of Sulphuric Acid, Sulphate of Soda, and Bleaching Powder. By JOHN LOMAS, Alkali Manufacturer, Newcastle-upon-Tyne and London. With 232 Illustrations and Working Drawings, and containing 390 pages of Text. Second Edition, with Additions. Super-royal 8vo, £1 10s. cloth.

"This book is written by a manufacturer for manufacturers. The working details of the most approved forms of apparatus are given, and these are accompanied by no less than 232 wood engravings, all of which may be used for the purposes of construction. Every step in the manufacture is very fully described in this manual, and each improvement explained."—*Athenæum.*

"The author is not one of those clever compilers who, on short notice, will 'read up' any conceivable subject, but a practical man in the best sense of the word. We find here not merely a sound and luminous explanation of the chemical principles of the trade, but a notice of numerous matters which have a most important bearing on the successful conduct of alkali works, but which are generally overlooked by even the most experienced technological authors."—*Chemical Review.*

Brewing.

A HANDBOOK FOR YOUNG BREWERS. By HERBERT EDWARDS WRIGHT, B.A. Crown 8vo, 3s. 6d. cloth.

"This little volume, containing such a large amount of good sense in so small a compass, ought to recommend itself to every brewery pupil."—*Brewers' Guardian.*

Commercial Chemical Analysis.

THE COMMERCIAL HANDBOOK OF CHEMICAL ANALYSIS; or, Practical Instructions for the determination of the Intrinsic or Commercial Value of Substances used in Manufactures, in Trades, and in the Arts. By A. NORMANDY, Editor of Rose's "Treatise on Chemical Analysis." New Edition, to a great extent Re-written by HENRY M. NOAD, Ph.D., F.R.S. With numerous Illustrations. Crown 8vo, 12s. 6d. cloth.

"We strongly recommend this book to our readers as a guide, alike indispensable to the housewife as to the pharmaceutical practitioner."—*Medical Times.*

"Essential to the analysts appointed under the new Act. The most recent results are given, and the work is well edited and carefully written."—*Nature.*

Dye-Wares and Colours.

THE MANUAL OF COLOURS AND DYE-WARES: Their Properties, Applications, Valuation, Impurities, and Sophistications. For the use of Dyers, Printers, Drysalters, Brokers, &c. By J. W. SLATER. Second Edition, Revised and greatly Enlarged. Crown 8vo, 7s. 6d. cloth.

"A complete encyclopædia of the *materia tinctoria*. The information given respecting each article is full and precise, and the methods of determining the value of articles such as these, so liable to sophistication, are given with clearness, and are practical as well as valuable."—*Chemist and Druggist.*

"There is no other work which covers precisely the same ground. To students preparing for examinations in dyeing and printing it will prove exceedingly useful."—*Chemical News.*

Pigments.

THE ARTIST'S MANUAL OF PIGMENTS. Showing their Composition, Conditions of Permanency, Non-Permanency, and Adulterations; Effects in Combination with Each Other and with Vehicles; and the most Reliable Tests of Purity. Together with the Science and Arts Department's Examination Questions on Painting. By H. C. STANDAGE. Second Edition, Revised. Small crown 8vo, 2s. 6d. cloth. [*Just published.*

"This work is indeed *multum-in-parvo*, and we can, with good conscience, recommend it to all who come in contact with pigments, whether as makers, dealers or users."—*Chemical Review.*

"This manual cannot fail to be a very valuable aid to all painters who wish their work to endure and be of a sound character; it is complete and comprehensive."—*Spectator.*

"The author supplies a great deal of very valuable information and memoranda as to the chemical qualities and artistic effect of the principal pigments used by painters."—*Builder.*

Gauging. Tables and Rules for Revenue Officers, Brewers, etc.

A POCKET BOOK OF MENSURATION AND GAUGING: Containing Tables, Rules and Memoranda for Revenue Officers, Brewers, Spirit Merchants, &c. By J. B. MANT (Inland Revenue). Oblong 18mo, 4s. leather, with elastic band. [*Just published.*

"This handy and useful book is adapted to the requirements of the Inland Revenue Department, and will be a favourite book of reference. The range of subjects is comprehensive, and the arrangement simple and clear."—*Civilian.*

"A most useful book. It should be in the hands of every practical brewer."—*Brewers' Journal.*

AGRICULTURE, FARMING, GARDENING, etc.

Agricultural Facts and Figures.

NOTE-BOOK OF AGRICULTURAL FACTS AND FIGURES FOR FARMERS AND FARM STUDENTS. By PRIMROSE MCCONNELL, Fellow of the Highland and Agricultural Society; late Professor of Agriculture, Glasgow Veterinary College. Third Edition. Royal 32mo, full roan, gilt edges, with elastic band, 4s.

"The most complete and comprehensive Note-book for Farmers and Farm Students that we have seen. It literally teems with information, and we can cordially recommend it to all connected with agriculture."—*North British Agriculturist.*

Youatt and Burn's Complete Grazier.

THE COMPLETE GRAZIER, and FARMER'S and CATTLE-BREEDER'S ASSISTANT. A Compendium of Husbandry; especially in the departments connected with the Breeding, Rearing, Feeding, and General Management of Stock; the Management of the Dairy, &c. With Directions for the Culture and Management of Grass Land, of Grain and Root Crops, the Arrangement of Farm Offices, the use of Implements and Machines, and on Draining, Irrigation, Warping, &c.; and the Application and Relative Value of Manures. By WILLIAM YOUATT, Esq., V.S. Twelfth Edition, Enlarged by ROBERT SCOTT BURN, Author of "Outlines of Modern Farming," "Systematic Small Farming," &c. One large 8vo volume, 860 pp., with 244 Illustrations, £1 1s. half-bound.

"The standard and text-book with the farmer and grazier."—*Farmers Magazine.*

"A treatise which will remain a standard work on the subject as long as British agriculture endures."—*Mark Lane Express* (First Notice).

"The book deals with all departments of agriculture, and contains an immense amount of valuable information. It is, in fact, an encyclopædia of agriculture put into readable form, and it is the only work equally comprehensive brought down to present date. It is excellently printed on thick paper, and strongly bound, and deserves a place in the library of every agriculturist."—*Mark Lane Express* (Second Notice).

36   CROSBY LOCKWOOD & SON'S CATALOGUE.

Flour Manufacture, Milling, etc.

FLOUR MANUFACTURE: A Treatise on Milling Science and Practice. By FRIEDRICH KICK, Imperial Regierungsrath, Professor of Mechanical Technology in the Imperial German Polytechnic Institute, Prague. Translated from the Second Enlarged and Revised Edition with Supplement. By H. H. P. POWLES, Assoc. Memb. Institution of Civil Engineers. Nearly 400 pp. Illustrated with 28 Folding Plates, and 167 Woodcuts. Royal 8vo, 25s. cloth. [*Just published.*

"This valuable work is, and will remain, the standard authority on the science of milling. . The miller who has read and digested this work will have laid the foundation, so to speak, of a successful career; he will have acquired a number of general principles which he can proceed to apply. In this handsome volume we at last have the accepted text-book of modern milling in good, sound English, which has little, if any, trace of the German idiom."—*The Miller.*

"Professor Kick treats the subject so thoroughly both on its theoretical and practical sides that his work is well suited to be a text-book of technical education anywhere."—*Scotsman.*

"The appearance of this celebrated work in English is very opportune, and British millers will, we are sure, not be slow in availing themselves of its pages."—*Millers' Gazette.*

Small Farming.

SYSTEMATIC SMALL FARMING; or, The Lessons of my Farm. Being an Introduction to Modern Farm Practice for Small Farmers in the Culture of Crops; The Feeding of Cattle; The Management of the Dairy, Poultry and Pigs; The Keeping of Farm Work Records; The Ensilage System, Construction of Silos, and other Farm Buildings; The Improvement of Neglected Farms, &c. By ROBERT SCOTT BURN, Author of "Outlines of Landed Estates' Management," and "Outlines of Farm Management," and Editor of "The Complete Grazier." With numerous Illustrations, crown 8vo, 6s. cloth.

"This is the completest book of its class we have seen, and one which every amateur farmer will read with pleasure and accept as a guide."—*Field.*

"Mr. Scott Burn's pages are severely practical, and the tone of the practical man is felt throughout. The book can only prove a treasure of aid and suggestion to the small farmer of intelligence and energy."—*British Quarterly Review.*

"The volume contains a vast amount of useful information. No branch of farming is left untouched, from the labour to be done to the results achieved."—*Glasgow Herald.*

Modern Farming.

OUTLINES OF MODERN FARMING. By R. SCOTT BURN. Soils, Manures, and Crops—Farming and Farming Economy—Cattle, Sheep, and Horses—Management of the Dairy, Pigs and Poultry—Utilisation of Town-Sewage, Irrigation, &c. Sixth Edition. In One Vol., 1,250 pp., half-bound, profusely Illustrated, 12s.

"The aim of the author has been to make his work at once comprehensive and trustworthy, and in this aim he has succeeded to a degree which entitles him to much credit."—*Morning Advertiser.*

"Eminently calculated to enlighten the agricultural community on the varied subjects of which it treats, and hence it should find a place in every farmer's library."—*City Press.*

"No farmer should be without this book."—*Banbury Guardian.*

Agricultural Engineering.

FARM ENGINEERING, THE COMPLETE TEXT-BOOK OF. Comprising Draining and Embanking; Irrigation and Water Supply; Farm Roads, Fences, and Gates; Farm Buildings, their Arrangement and Construction, with Plans and Estimates; Barn Implements and Machines; Field Implements and Machines; Agricultural Surveying, Levelling, &c. By Prof. JOHN SCOTT, Editor of the *Farmers' Gazette*, late Professor of Agriculture and Rural Economy at the Royal Agricultural College, Cirencester, &c. &c. In One Vol., 1,150 pages, half-bound, with over 600 Illustrations, 12s.

"Written with great care, as well as with knowledge and ability. The author has done his work well; we have found him a very trustworthy guide wherever we have tested his statements. The volume will be of great value to agricultural students."—*Mark Lane Express.*

"For a young agriculturist we know of no handy volume so likely to be more usefully studied."—*Bell's Weekly Messenger.*

English Agriculture.

THE FIELDS OF GREAT BRITAIN: A Text-Book of Agriculture, adapted to the Syllabus of the Science and Art Department. For Elementary and Advanced Students. By HUGH CLEMENTS (Board of Trade). 18mo, 2s. 6d. cloth.

"A most comprehensive volume, giving a mass of information."—*Agricultural Economist.*

"It is a long time since we have seen a book which has pleased us more, or which contains a vast and useful fund of knowledge."—*Educational Times.*

AGRICULTURE, FARMING, GARDENING, etc. 37

Farm and Estate Book-keeping.
BOOK-KEEPING FOR FARMERS & ESTATE OWNERS. A Practical Treatise, presenting, in Three Plans, a System adapted to all Classes of Farms. By JOHNSON M. WOODMAN, Chartered Accountant. Crown 8vo, 3s. 6d. cloth.

"Will be found of great assistance by those who intend to commence a system of book-keeping, the author's examples being clear and explicit, and his explanations, while full and accurate, being to a large extent free from technicalities."—*Live Stock Journal*.

Farm Account Book.
WOODMAN'S YEARLY FARM ACCOUNT BOOK. Giving a Weekly Labour Account and Diary, and showing the Income and Expenditure under each Department of Crops, Live Stock, Dairy, &c. &c. With Valuation, Profit and Loss Account, and Balance Sheet at the end of the Year, and an Appendix of Forms. Ruled and Headed for Entering a Complete Record of the Farming Operations. By JOHNSON M. WOODMAN, Chartered Accountant, Author of "Book-keeping for Farmers." Folio, 7s. 6d. half-bound.

"Contains every requisite orm for keeping farm accounts readily and accurately."—*Agriculture*.

Early Fruits, Flowers and Vegetables.
THE FORCING GARDEN; or, How to Grow Early Fruits, Flowers, and Vegetables. With Plans, and Estimates for Building Glasshouses, Pits and Frames. Containing also Original Plans for Double Glazing, a New Method of Growing the Gooseberry under Glass, &c. &c., and on Ventilation, Protecting Vine Borders, &c. With Illustrations. By SAMUEL WOOD. Crown 8vo, 3s. 6d. cloth.

"Mr. Wood's book is an original and exhaustive answer to the question 'How to Grow Early Fruits, Flowers and Vegetables?'"—*Land and Water*.

Good Gardening.
A PLAIN GUIDE TO GOOD GARDENING; or, How to Grow Vegetables, Fruits, and Flowers. With Practical Notes on Soils, Manures, Seeds, Planting, Laying-out of Gardens and Grounds, &c. By S. WOOD. Third Edition, with considerable Additions, &c., and numerous Illustrations. Crown 8vo, 5s. cloth.

"A very good book, and one to be highly recommended as a practical guide."—*Athenæum*.
"May be recommended to young gardeners, cottagers, and specially to amateurs, for the plain and trustworthy information it gives on common matters."—*Gardeners' Chronicle*.

Gainful Gardening.
MULTUM-IN-PARVO GARDENING; or, How to make One Acre of Land produce £620 a-year by the Cultivation of Fruits and Vegetables; also, How to Grow Flowers in Three Glass Houses, so as to realise £176 per annum clear Profit. By SAMUEL WOOD, Author of "Good Gardening," &c. Fourth and cheaper Edition, Revised, with Additions. Crown 8vo, 1s. sewed.

"We are bound to recommend it as not only suited to the case of the amateur and gentleman's gardener, but to the market grower."—*Gardeners' Magazine*.

Gardening for Ladies.
THE LADIES' MULTUM-IN-PARVO FLOWER GARDEN, and Amateurs' Complete Guide. With Illustrations. By SAMUEL WOOD. Crown 8vo, 3s. 6d. cloth.

"Full of shrewd hints and useful instructions, based a lifetime of experience."—*Scotsman*.

Receipts for Gardeners.
GARDEN RECEIPTS. Edited by CHARLES W. QUIN. 12mo, 1s. 6d. cloth limp.

"A useful and handy book, containing a good deal of valuable information."—*Athenæum*.

Kitchen Gardening.
THE KITCHEN AND MARKET GARDEN. By Contributors to "The Garden." Compiled by C. W. SHAW. 12mo, 3s. 6d. cloth boards.

"The most valuable compendium of kitchen and market-garden work published."—*Farmer*

Cottage Gardening.
COTTAGE GARDENING; or, Flowers, Fruits, and Vegetables for Small Gardens. By E. HOBDAY. 12mo, 1s. 6d. cloth limp.

"Contains much usefu information at a small charge."—*Glasgow Herald*.

ESTATE MANAGEMENT, AUCTIONEERING, LAW, etc.

Hudson's Land Valuer's Pocket-Book.

THE LAND VALUER'S BEST ASSISTANT: Being Tables on a very much Improved Plan, for Calculating the Value of Estates. With Tables for reducing Scotch, Irish, and Provincial Customary Acres to Statute Measure, &c. By R. HUDSON, C.E. New Edition. Royal 32mo, leather, elastic band, 4s.

"This new edition includes tables for ascertaining the value of leases for any term of years; and for showing how to lay out plots of ground of certain acres in forms, square, round, &c., with valuable rules for ascertaining the probable worth of standing timber to any amount; and is of incalculable value to the country gentleman and professional man."—*Farmers' Journal.*

Ewart's Land Improver's Pocket-Book.

THE LAND IMPROVER'S POCKET-BOOK OF FORMULÆ, TABLES and MEMORANDA *required in any Computation relating to the Permanent Improvement of Landed Property.* By JOHN EWART, Land Surveyor and Agricultural Engineer. Second Edition, Revised. Royal 32mo, oblong, leather, gilt edges, with elastic band, 4s.

"A compendious and handy little volume."—*Spectator.*

Complete Agricultural Surveyor's Pocket-Book.

THE LAND VALUER'S AND LAND IMPROVER'S COMPLETE POCKET-BOOK. Consisting of the above Two Works bound together. Leather, gilt edges, with strap, 7s. 6d.

"Hudson's book is the best ready-reckoner on matters relating to the valuation of land and crops, and its combination with Mr. Ewart's work greatly enhances the value and usefulness of the latter-mentioned. . . . It is most useful as a manual for reference."—*North of England Farmer.*

Auctioneer's Assistant.

THE APPRAISER, AUCTIONEER, BROKER, HOUSE AND ESTATE AGENT AND VALUER'S POCKET ASSISTANT, for the Valuation for Purchase, Sale, or Renewal of Leases, Annuities and Reversions, and of property generally; with Prices for Inventories, &c. By JOHN WHEELER, Valuer, &c. Fifth Edition, re-written and greatly extended by C. NORRIS, Surveyor, Valuer, &c. Royal 32mo, 5s. cloth.

"A neat and concise book of reference, containing an admirable and clearly-arranged list of prices for inventories, and a very practical guide to determine the value of furniture, &c."—*Standard.*

"Contains a large quantity of varied and useful information as to the valuation for purchase, sale, or renewal of leases, annuities and reversions, and of property generally, with prices for inventories, and a guide to determine the value of interior fittings and other effects."—*Builder.*

Auctioneering.

AUCTIONEERS: *Their Duties and Liabilities.* By ROBERT SQUIBBS, Auctioneer. Demy 8vo, 10s. 6d. cloth.

"The position and duties of auctioneers treated compendiously and clearly."—*Builder.*

"Every auctioneer ought to possess a copy of this excellent work."—*Ironmonger.*

"Of great value to the profession. . . . We readily welcome this book from the fact that it treats the subject in a manner somewhat new to the profession."—*Estates Gazette.*

Legal Guide for Pawnbrokers.

THE PAWNBROKERS', FACTORS' AND MERCHANTS' GUIDE TO THE LAW OF LOANS AND PLEDGES. With the Statutes and a Digest of Cases on Rights and Liabilities, Civil and Criminal, as to Loans and Pledges of Goods, Debentures, Mercantile and other Securities. By H. C. FOLKARD, Esq., Barrister-at-Law, Author of "The Law of Slander and Libel," &c. With Additions and Corrections to 1876. Fcap. 8vo, 3s. 6d. cloth.

"This work contains simply everything that requires to be known concerning the department of the law of which it treats. We can safely commend the book as unique and very nearly perfect.'—*Iron.*

"The task undertaken by Mr. Folkard has been very satisfactorily performed. . . . Such explanations as are needful have been supplied with great clearness and with due regard to brevity. *City Press.*

ESTATE MANAGEMENT, AUCTIONEERING, LAW, etc. 39

How to Invest.
HINTS FOR INVESTORS: Being an Explanation of the Mode of Transacting Business on the Stock Exchange. To which are added Comments on the Fluctuations and Table of Quarterly Average prices of Consols since 1759. Also a Copy of the London Daily Stock and Share List. By WALTER M. PLAYFORD, Sworn Broker. Crown 8vo, 2s. cloth.

"An invaluable guide to investors and speculators."—*Bullionist.*

Metropolitan Rating Appeals.
REPORTS OF APPEALS HEARD BEFORE THE COURT OF GENERAL ASSESSMENT SESSIONS, from the Year 1871 to 1885. By EDWARD RYDE and ARTHUR LYON RYDE. Fourth Edition, brought down to the Present Date, with an Introduction to the Valuation (Metropolis) Act, 1869, and an Appendix by WALTER C. RYDE, of the Inner Temple, Barrister-at-Law. 8vo, 16s. cloth.

"A useful work, occupying a place mid-way between a handbook for a lawyer and a guide to the surveyor. It is compiled by a gentleman eminent in his profession as a land agent, whose speciality, it is acknowledged, lies in the direction of assessing property for rating purposes."—*Land Agents' Record.*

House Property.
HANDBOOK OF HOUSE PROPERTY. A Popular and Practical Guide to the Purchase, Mortgage, Tenancy, and Compulsory Sale of Houses and Land, including the Law of Dilapidations and Fixtures; with Examples of all kinds of Valuations, Useful Information on Buildings, and Suggestive Elucidations of Fine Art. By E. L. TARBUCK, Architect and Surveyor. Fourth Edition, 12mo, 5s. cloth. [*Just published.*

"The advice is thoroughly practical."—*Law Journal.*
"This is a well-written and thoughtful work. We commend the work to the careful study of all interested in questions affecting houses and land."—*Land Agents' Record* (First Notice).
"Carefully brought up to date, and much improved by the addition of a division on fine art."
—*Land Agents' Record* (Second Notice).

Inwood's Estate Tables.
TABLES FOR THE PURCHASING OF ESTATES, Freehold, Copyhold, or Leasehold; Annuities, Advowsons, etc., and for the Renewing of Leases held under Cathedral Churches, Colleges, or other Corporate bodies, for Terms of Years certain, and for Lives; also for Valuing Reversionary Estates, Deferred Annuities, Next Presentations, &c.; together with SMART'S Five Tables of Compound Interest, and an Extension of the same to Lower and Intermediate Rates. By W. INWOOD. 22nd Edition, with considerable Additions, and new and valuable Tables of Logarithms for the more Difficult Computations of the Interest of Money, Discount, Annuities, &c., by M. FEDOR THOMAN, of the Société Crédit Mobilier of Paris. 12mo, 8s. cloth.

"Those interested in the purchase and sale of estates, and in the adjustment of compensation cases, as well as in transactions in annuities, life insurances, &c., will find the present edition of eminent service."—*Engineering.*
"'Inwood's Tables' still maintain a most enviable reputation. The new issue has been enriched by large additional contributions by M. Fedor Theman, whose carefully arranged Tables cannot fail to be of the utmost utility."—*Mining Journal.*

Agricultural and Tenant-Right Valuation.
THE AGRICULTURAL AND TENANT-RIGHT-VALUER'S ASSISTANT. A Practical Handbook on Measuring and Estimating the Contents, Weights and Values of Agricultural Produce and Timber, the Values of Estates and Agricultural Labour, Forms of Tenant-Right-Valuations, Scales of Compensation under the Agricultural Holdings Act, 1883, &c. &c. By TOM BRIGHT, Agricultural Surveyor, Author of "The Live Stock of North Devon," &c. Crown 8vo, 3s. 6d. cloth. [*Just published.*

"Full of tables and examples in connection with the valuation of tenant-right, estates, labour, contents, and weights of timber, and farm produce of all kinds. The book is well calculated to assist the valuer in the discharge of his duty."—*Agricultural Gazette.*
"An eminently practical handbook, full of practical tables and data of undoubted interest and value to surveyors and auctioneers in preparing valuations of all kinds."—*Farmer.*
"Shows at a glance the value of land, crops, and the cost of seeding, harvesting, &c. &c. It is a really practical and useful handbook, for which we anticipate a very large sale."—*Reading Mercury.*

A Complete Epitome of the Laws of this Country.

EVERY MAN'S OWN LAWYER: A Handy-book of the Principles of Law and Equity. By A BARRISTER. Twenty-fifth Edition. Reconstructed, Thoroughly Revised, and much Enlarged. Brought down to the end of last Session, and including careful Digests of—*Coroners Act*, 1887; *Probation of First Offenders Act*, 1887; *Margarine Act*, 1887; *Agricultural Holdings (England) Act*, 1883; *Cottage Gardens (Compensation for Crops) Act*, 1887; *Bankruptcy Act*, 1883; *Allotments Act*, 1887; *Merchandise Trade Marks Act*, 1887; *Truck Amendment Act*, 1887; *Water Companies (Regulation of Powers) Act*, 1887; *Registration of Deeds of Arrangements Act*, 1887. Crown 8vo, 684 pp., price 6s. 8d. (saved at every consultation!), strongly bound in cloth. [*Just published.*

*** THE BOOK WILL BE FOUND TO COMPRISE (AMONGST OTHER MATTER)—

THE RIGHTS AND WRONGS OF INDIVIDUALS—MERCANTILE AND COMMERCIAL LAW—PARTNERSHIPS, CONTRACTS AND AGREEMENTS—GUARANTEES, PRINCIPALS AND AGENTS—CRIMINAL LAW—PARISH LAW—COUNTY COURT LAW—GAME AND FISHERY LAWS—POOR MEN'S LAWSUITS—LAWS OF BANKRUPTCY—WAGERS—CHEQUES, BILLS AND NOTES—COPYRIGHT—ELECTIONS AND REGISTRATION—INSURANCE—LIBEL AND SLANDER—MARRIAGE AND DIVORCE—MERCHANT SHIPPING—MORTGAGES—SETTLEMENTS—STOCK EXCHANGE PRACTICE—TRADE MARKS AND PATENTS—TRESPASS—NUISANCES—TRANSFER OF LAND—WILLS, &c. &c. Also LAW FOR LANDLORD AND TENANT—MASTER AND SERVANT—HEIRS—DEVISEES AND LEGATEES—HUSBAND AND WIFE—EXECUTORS AND TRUSTEES—GUARDIAN AND WARD—MARRIED WOMEN AND INFANTS—LENDER, BORROWER AND SURETIES—DEBTOR AND CREDITOR—PURCHASER AND VENDOR—COMPANIES—FRIENDLY SOCIETIES—CLERGYMEN—CHURCHWARDENS—MEDICAL PRACTITIONERS—BANKERS—FARMERS—CONTRACTORS—STOCK BROKERS—SPORTSMEN—GAMEKEEPERS—FARRIERS—HORSE DEALERS—AUCTIONEERS—HOUSE AGENTS—INNKEEPERS—BAKERS—MILLERS—PAWNBROKERS—SURVEYORS—RAILWAYS AND CARRIBRS—CONSTABLES—SEAMEN—SOLDIERS, &c. &c.

☞ *The following subjects may be mentioned as some of those which have received special attention during the present revision:*—Marriage of British Subjects Abroad; Police Constables; Pawnbrokers; Intoxicating Liquors Licensing; Domestic Servants; Landlord and Tenant; Vendors and Purchasers; Parliamentary Elections; Municipal Elections; Local Elections; Corrupt Practices at all Elections; Public Health and Local Government and Nuisances; Highways; Churchwardens; Vestry Meetings; Rates.

It is believed that the extensions and amplifications of the present edition, while intended to meet the requirements of the ordinary Englishman, will also have the effect of rendering the book useful to the legal practitioner in the country.

One result of the reconstruction and revision, with the extensive additions thereby necessitated, has been *the enlargement of the book by nearly a hundred and fifty pages*, while the price remains as before.

*** OPINIONS OF THE PRESS.

"It is a complete code of English Law, written in plain language, which all can understand. . . . Should be in the hands of every business man, and all who wish to abolish lawyers' bills."—*Weekly Times.*

"A useful and concise epitome of the law, compiled with considerable care. —*Law Magazine.*

"A concise, cheap and complete epitome of the English law. So plainly written that he who runs may read, and he who reads may understand."—*Figaro.*

"A dictionary of legal facts well put together. The book is a very useful one."—*Spectator.*

"A work which has long been wanted, which is thoroughly well done, and which we most cordially recommend."—*Sunday Times.*

Private Bill Legislation and Provisional Orders.

HANDBOOK FOR THE USE OF SOLICITORS AND ENGINEERS Engaged in Promoting Private Acts of Parliament and Provisional Orders, for the Authorization of Railways, Tramways, Works for the Supply of Gas and Water, and other undertakings of a like character. By L. LIVINGSTON MACASSEY, of the Middle Temple, Barrister-at-Law, and Member of the Institution of Civil Engineers; Author of "Hints on Water Supply." Demy 8vo, 950 pp., price 25s. cloth. [*Just published.*

"The volume is a desideratum on a subject which can be only acquired by practical experience, and the order of procedure in Private Bill Legislation and Provisional Orders is followed. The author's suggestions and notes will be found of great value to engineers and others professionally engaged in this class of practice."—*Building News.*

"The author's double experience as an engineer and barrister has eminently qualified him for the task, and enabled him to approach the subject alike from an engineering and legal point of view. The volume will be found a great help both to engineers and lawyers engaged in promoting Private Acts of Parliament and Provisional Orders."—*Local Government Chronicle.*

J. OGDEN AND CO. LIMITED, PRINTERS, GREAT SAFFRON HILL, E.C.

Weale's Rudimentary Series.

LONDON, 1862.
THE PRIZE MEDAL
Was awarded to the Publishers of
"WEALE'S SERIES."

A NEW LIST OF
WEALE'S SERIES
RUDIMENTARY SCIENTIFIC, EDUCATIONAL, AND CLASSICAL.

Comprising nearly Three Hundred and Fifty distinct works in almost every department of Science, Art, and Education, recommended to the notice of Engineers, Architects, Builders, Artisans, and Students generally, as well as to those interested in Workmen's Libraries, Literary and Scientific Institutions, Colleges, Schools, Science Classes, &c., &c.

☞ "WEALE'S SERIES includes Text-Books on almost every branch of Science and Industry, comprising such subjects as Agriculture, Architecture and Building, Civil Engineering, Fine Arts, Mechanics and Mechanical Engineering, Physical and Chemical Science, and many miscellaneous Treatises. The whole are constantly undergoing revision, and new editions, brought up to the latest discoveries in scientific research, are constantly issued. The prices at which they are sold are as low as their excellence is assured."—*American Literary Gazette.*

"Amongst the literature of technical education, WEALE'S SERIES has ever enjoyed a high reputation, and the additions being made by Messrs. CROSBY LOCKWOOD & SON render the series even more complete, and bring the information upon the several subjects down to the present time."—*Mining Journal.*

"It is not too much to say that no books have ever proved more popular with, or more useful to, young engineers and others than the excellent treatises comprised in WEALE'S SERIES."—*Engineer.*

"The excellence of WEALE'S SERIES is now so well appreciated, that it would be wasting our space to enlarge upon their general usefulness and value."—*Builder.*

"WEALE'S SERIES has become a standard as well as an unrivalled collection of treatises in all branches of art and science."—*Public Opinion.*

PHILADELPHIA, 1876.
THE PRIZE MEDAL
Was awarded to the Publishers for
Books: Rudimentary, Scientific,
"WEALE'S SERIES," ETC.

CROSBY LOCKWOOD & SON,
7, STATIONERS' HALL COURT, LUDGATE HILL, LONDON, E.C.

WEALE'S RUDIMENTARY SCIENTIFIC SERIES.

⁎ The volumes of this Series are freely Illustrated with Woodcuts, or otherwise, where requisite. Throughout the following List it must be understood that the books are bound in limp cloth, unless otherwise stated; *but the volumes marked with a ‡ may also be had strongly bound in cloth boards for 6d. extra.*

N.B.—*In ordering from this List it is recommended, as a means of facilitating business and obviating error, to quote the numbers affixed to the volumes, as well as the titles and prices.*

CIVIL ENGINEERING, SURVEYING, ETC.

No.
31. **WELLS AND WELL-SINKING.** By JOHN GEO. SWINDELL, A.R.I.B.A., and G. R. BURNELL, C.E. Revised Edition. With a New Appendix on the Qualities of Water. Illustrated. 2s.
35. **THE BLASTING AND QUARRYING OF STONE**, for Building and other Purposes. By Gen. Sir J. BURGOYNE, Bart. 1s. 6d.
43. **TUBULAR, AND OTHER IRON GIRDER BRIDGES**, particularly describing the Britannia and Conway Tubular Bridges. By G. DRYSDALE DEMPSEY, C.E. Fourth Edition. 2s.
44. **FOUNDATIONS AND CONCRETE WORKS**, with Practical Remarks on Footings, Sand, Concrete, Béton, Pile-driving, Caissons, and Cofferdams, &c. By E. DOBSON. Fifth Edition. 1s. 6d.
60. **LAND AND ENGINEERING SURVEYING.** By T. BAKER, C.E. Fourteenth Edition, revised by Professor J. R. YOUNG. 2s.‡
80*. **EMBANKING LANDS FROM THE SEA.** With examples and Particulars of actual Embankments, &c. By J. WIGGINS, F.G.S. 2s.
81. **WATER WORKS**, for the Supply of Cities and Towns. With a Description of the Principal Geological Formations of England as influencing Supplies of Water, &c. By S. HUGHES, C.E. New Edition. 4s.‡
118. **CIVIL ENGINEERING IN NORTH AMERICA**, a Sketch of. By DAVID STEVENSON, F.R.S.E., &c. Plates and Diagrams. 3s.
167. **IRON BRIDGES, GIRDERS, ROOFS, AND OTHER WORKS.** By FRANCIS CAMPIN, C.E. 2s. 6d.‡
197. **ROADS AND STREETS.** By H. LAW, C.E., revised and enlarged by D. K. CLARK, C.E., including pavements of Stone, Wood, Asphalte, &c. 4s. 6d.‡
203. **SANITARY WORK IN THE SMALLER TOWNS AND IN VILLAGES.** By C. SLAGG, A.M.I.C.E. Revised Edition. 3s.‡
212. **GAS-WORKS, THEIR CONSTRUCTION AND ARRANGEMENT**; and the Manufacture and Distribution of Coal Gas. Originally written by SAMUEL HUGHES, C.E. Re-written and enlarged by WILLIAM RICHARDS, C.E. Seventh Edition, with important additions. 5s. 6d.‡
213. **PIONEER ENGINEERING.** A Treatise on the Engineering Operations connected with the Settlement of Waste Lands in New Countries. By EDWARD DOBSON, Assoc. Inst. C.E. 4s. 6d.‡
216. **MATERIALS AND CONSTRUCTION;** A Theoretical and Practical Treatise on the Strains, Designing, and Erection of Works of Construction. By FRANCIS CAMPIN, C.E. Second Edition, revised. 3s.‡
219. **CIVIL ENGINEERING.** By HENRY LAW, M.Inst. C.E. Including HYDRAULIC ENGINEERING by GEO. R. BURNELL, M.Inst. C.E. Seventh Edition, revised, with large additions by D. KINNEAR CLARK, M.Inst. C.E. 6s. 6d., Cloth boards, 7s. 6d.
268. **THE DRAINAGE OF LANDS, TOWNS, & BUILDINGS.** By G. D. DEMPSEY, C.E. Revised, with large Additions on Recent Practice in Drainage Engineering, by D. KINNEAR CLARK, M.I.C.E. Second Edition, Corrected. 4s. 6d.‡ [*Just published.*

☞ *The ‡ indicates that these vols. may be had strongly bound at 6d. extra.*

LONDON : CROSBY LOCKWOOD AND SON.

MECHANICAL ENGINEERING, ETC.

33. *CRANES*, the Construction of, and other Machinery for Raising Heavy Bodies. By JOSEPH GLYNN, F.R.S. Illustrated. 1s. 6d.
34. *THE STEAM ENGINE*. By Dr. LARDNER. Illustrated. 1s. 6d.
59. *STEAM BOILERS:* their Construction and Management. By R. ARMSTRONG, C.E. Illustrated. 1s. 6d.
82. *THE POWER OF WATER*, as applied to drive Flour Mills, and to give motion to Turbines, &c. By JOSEPH GLYNN, F.R.S. 2s.‡
98. *PRACTICAL MECHANISM*, the Elements of; and Machine Tools. By T. BAKER, C.E. With Additions by J. NASMYTH, C.E. 2s. 6d.‡
139. *THE STEAM ENGINE*, a Treatise on the Mathematical Theory of, with Rules and Examples for Practical Men. By T. BAKER, C.E. 1s. 6d.
164. *MODERN WORKSHOP PRACTICE*, as applied to Steam Engines, Bridges, Ship-building, Cranes, &c. By J. G. WINTON. Fourth Edition, much enlarged and carefully revised. 3s. 6d.‡ [*Just published.*
165. *IRON AND HEAT*, exhibiting the Principles concerned in the Construction of Iron Beams, Pillars, and Girders. By J. ARMOUR. 2s. 6d.‡
166. *POWER IN MOTION:* Horse-Power, Toothed-Wheel Gearing, Long and Short Driving Bands, and Angular Forces. By J. ARMOUR, 2s.‡
171. *THE WORKMAN'S MANUAL OF ENGINEERING DRAWING*. By J. MAXTON. 6th Edn. With 7 Plates and 350 Cuts. 3s. 6d.‡
190. *STEAM AND THE STEAM ENGINE*, Stationary and Portable. Being an Extension of the Elementary Treatise on the Steam Engine of Mr. JOHN SEWELL. By D. K. CLARK, M.I.C.E. 3s. 6d.‡
200. *FUEL*, its Combustion and Economy. By C. W. WILLIAMS. With Recent Practice in the Combustion and Economy of Fuel—Coal, Coke, Wood, Peat, Petroleum, &c.—by D. K. CLARK, M.I.C.E. 3s. 6d.‡
202. *LOCOMOTIVE ENGINES*. By G. D. DEMPSEY, C.E.; with large additions by D. KINNEAR CLARK, M.I.C.E. 3s.‡
211. *THE BOILERMAKER'S ASSISTANT* in Drawing, Templating, and Calculating Boiler and Tank Work. By JOHN COURTNEY. Practical Boiler Maker. Edited by D. K. CLARK, C.E. 100 Illustrations. 2s.
217. *SEWING MACHINERY:* Its Construction, History, &c., with full Technical Directions for Adjusting, &c. By J. W. URQUHART, C.E. 2s.‡
223. *MECHANICAL ENGINEERING*. Comprising Metallurgy, Moulding, Casting, Forging, Tools, Workshop Machinery, Manufacture of the Steam Engine, &c. By FRANCIS CAMPIN, C.E. Second Edition. 2s. 6d.‡
236. *DETAILS OF MACHINERY*. Comprising Instructions for the Execution of various Works in Iron. By FRANCIS CAMPIN, C.E. 3s.‡
237. *THE SMITHY AND FORGE;* including the Farrier's Art and Coach Smithing. By W. J. E. CRANE. Illustrated. 2s. 6d.‡
238. *THE SHEET-METAL WORKER'S GUIDE;* a Practical Handbook for Tinsmiths, Coppersmiths, Zincworkers, &c. With 94 Diagrams and Working Patterns. By W. J. E. CRANE. Second Edition, revised. 1s. 6d.
251. *STEAM AND MACHINERY MANAGEMENT:* with Hints on Construction and Selection. By M. POWIS BALE, M.I.M.E. 2s. 6d.‡
254. *THE BOILERMAKER'S READY-RECKONER*. By J. COURTNEY. Edited by D. K. CLARK, C.E. 4s., limp; 5s., half-bound.
255. *LOCOMOTIVE ENGINE-DRIVING*. A Practical Manual for Engineers in charge of Locomotive Engines. By MICHAEL REYNOLDS, M.S.E. Eighth Edition. 3s. 6d., limp: 4s. 6d. cloth boards.
256. *STATIONARY ENGINE-DRIVING*. A Practical Manual Engineers in charge of Stationary Engines. By MICHAEL REYNOLDS, M.S.E. Third Edition. 3s. 6d. limp; 4s. 6d. cloth boards.
260. *IRON BRIDGES OF MODERATE SPAN:* their Construction and Erection. By HAMILTON W. PENDRED, C.E. 2s.

☞ *The ‡ indicates that these vols. may be had strongly bound at 6d. extra.*

7, STATIONERS' HALL COURT, LUDGATE HILL, E.C.

MINING, METALLURGY, ETC.

4. *MINERALOGY,* Rudiments of; a concise View of the General Properties of Minerals. By A. RAMSAY, F.G.S., F.R.G.S., &c. Third Edition, revised and enlarged. Illustrated. 3s. 6d.‡

117. *SUBTERRANEOUS SURVEYING,* with and without the Magnetic Needle. By T. FENWICK and T. BAKER, C.E. Illustrated. 2s. 6d.‡

133. *METALLURGY OF COPPER.* By R. H. LAMBORN. 2s. 6d.‡

135. *ELECTRO-METALLURGY;* Practically Treated. By ALEXANDER WATT. Ninth Edition, enlarged and revised, with additional Illustrations, and including the most recent Processes. 3s. 6d.‡

172. *MINING TOOLS,* Manual of. For the Use of Mine Managers, Agents, Students, &c. By WILLIAM MORGANS. 2s. 6d.

172*. *MINING TOOLS, ATLAS* of Engravings to Illustrate the above, containing 235 Illustrations, drawn to Scale. 4to. 4s. 6d.

176. *METALLURGY OF IRON.* Containing History of Iron Manufacture, Methods of Assay, and Analyses of Iron Ores, Processes of Manufacture of Iron and Steel, &c. By H. BAUERMAN, F.G.S. Sixth Edition, revised and enlarged. 5s.‡ [*Just published.*

180. *COAL AND COAL MINING.* By SIR WARINGTON W. SMYTH, M.A., F.R.S. Seventh Edition, revised. 3s. 6d.‡ [*Just published.*

195. *THE MINERAL SURVEYOR AND VALUER'S COMPLETE GUIDE.* By W. LINTERN, Mining Engineer. Third Edition, with an Appendix on Magnetic and Angular Surveying. With Four Plates. 3s. 6d.‡ [*Just published.*

214. *SLATE AND SLATE QUARRYING,* Scientific, Practical, and Commercial. By D. C. DAVIES, F.G.S., Mining Engineer, &c. 3s.‡

264. *A FIRST BOOK OF MINING AND QUARRYING,* with the Sciences connected therewith, for Primary Schools and Self Instruction. By J. H. COLLINS, F.G.S. Second Edition, with additions. 1s. 6d.

ARCHITECTURE, BUILDING, ETC.

16. *ARCHITECTURE—ORDERS*—The Orders and their Æsthetic Principles. By W. H. LEEDS. Illustrated. 1s. 6d.

17. *ARCHITECTURE—STYLES*—The History and Description of the Styles of Architecture of Various Countries, from the Earliest to the Present Period. By T. TALBOT BURY, F.R.I.B.A., &c. Illustrated. 2s.
*** ORDERS AND STYLES OF ARCHITECTURE, *in One Vol.,* 3s. 6d.

18. *ARCHITECTURE—DESIGN*—The Principles of Design in Architecture, as deducible from Nature and exemplified in the Works of the Greek and Gothic Architects. By E. L. GARBETT, Architect. Illustrated. 2s. 6d.
*** *The three preceding Works, in One handsome Vol., half bound, entitled* "MODERN ARCHITECTURE," *price* 6s.

22. *THE ART OF BUILDING,* Rudiments of. General Principles of Construction, Materials used in Building, Strength and Use of Materials, Working Drawings, Specifications, and Estimates. By E. DOBSON, 2s.‡

25. *MASONRY AND STONECUTTING:* Rudimentary Treatise on the Principles of Masonic Projection and their application to Construction. By EDWARD DOBSON, M.R.I.B.A., &c. 2s. 6d.‡

42. *COTTAGE BUILDING.* By C. BRUCE ALLEN, Architect. Tenth Edition, revised and enlarged. With a Chapter on Economic Cottages for Allotments, by EDWARD E. ALLEN, C.E. 2s.

45. *LIMES, CEMENTS, MORTARS, CONCRETES, MASTICS,* PLASTERING, &c. By G. R. BURNELL, C.E. Thirteenth Edition. 1s. 6d.

57. *WARMING AND VENTILATION.* An Exposition of the General Principles as applied to Domestic and Public Buildings, Mines, Lighthouses, Ships, &c. By C. TOMLINSON, F.R.S., &c. Illustrated. 3s.

☞ *The* ‡ *indicates that these vols. may be had strongly bound at* 6d. *extra.*

LONDON : CROSBY LOCKWOOD AND SON,

Architecture, Building, etc., *continued*.

111. *ARCHES, PIERS, BUTTRESSES, &c.*: Experimental Essays on the Principles of Construction. By W. BLAND. Illustrated. 1s. 6d.
116. *THE ACOUSTICS OF PUBLIC BUILDINGS*; or, The Principles of the Science of Sound applied to the purposes of the Architect and Builder. By T. ROGER SMITH, M.R.I.B.A., Architect. Illustrated. 1s. 6d.
127. *ARCHITECTURAL MODELLING IN PAPER*, the Art of. By T. A. RICHARDSON, Architect. Illustrated. 1s. 6d.
128. *VITRUVIUS — THE ARCHITECTURE OF MARCUS VITRUVIUS POLLO.* In Ten Books. Translated from the Latin by JOSEPH GWILT, F.S.A., F.R.A.S. With 23 Plates. 5s.
130. *GRECIAN ARCHITECTURE*, An Inquiry into the Principles of Beauty in; with an Historical View of the Rise and Progress of the Art in Greece. By the EARL OF ABERDEEN. 1s.

⁂ *The two preceding Works in One handsome Vol., half bound, entitled* "ANCIENT ARCHITECTURE," *price* 6s.

132. *THE ERECTION OF DWELLING-HOUSES.* Illustrated by a Perspective View, Plans, Elevations, and Sections of a pair of Semi-detached Villas, with the Specification, Quantities, and Estimates, &c. By S. H. BROOKS. New Edition, with Plates. 2s. 6d.‡
156. *QUANTITIES & MEASUREMENTS* in Bricklayers', Masons', Plasterers', Plumbers', Painters', Paperhangers', Gilders', Smiths', Carpenters' and Joiners' Work. By A. C. BEATON, Surveyor. New Edition. 1s. 6d.
175. *LOCKWOOD'S BUILDER'S PRICE BOOK FOR* 1890. A Comprehensive Handbook of the Latest Prices and Data for Builders, Architects, Engineers, and Contractors. Re-constructed, Re-written, and greatly Enlarged. By FRANCIS T. W. MILLER, A.R.I.B.A. 640 pages. 3s. 6d.‡ [*Just published.*
182. *CARPENTRY AND JOINERY* — THE ELEMENTARY PRINCIPLES OF CARPENTRY. Chiefly composed from the Standard Work of THOMAS TREDGOLD, C.E. With a TREATISE ON JOINERY by E. WYNDHAM TARN, M.A. Fourth Edition, Revised. 3s. 6d.‡
182*. *CARPENTRY AND JOINERY. ATLAS* of 35 Plates to accompany the above. With Descriptive Letterpress. 4to. 6s.
185. *THE COMPLETE MEASURER*; the Measurement of Boards, Glass, &c.; Unequal-sided, Square-sided, Octagonal-sided, Round Timber and Stone, and Standing Timber, &c. By RICHARD HORTON. Fifth Edition. 4s.; strongly bound in leather, 5s.
187. *HINTS TO YOUNG ARCHITECTS.* By G. WIGHTWICK. New Edition. By G. H. GUILLAUME. Illustrated. 3s. 6d.‡
188. *HOUSE PAINTING, GRAINING, MARBLING, AND SIGN WRITING*: with a Course of Elementary Drawing for House-Painters, Sign-Writers, &c., and a Collection of Useful Receipts. By ELLIS A. DAVIDSON. Fifth Edition. With Coloured Plates. 5s. cloth limp; 6s. cloth boards.
189. *THE RUDIMENTS OF PRACTICAL BRICKLAYING.* In Six Sections: General Principles; Arch Drawing, Cutting, and Setting; Pointing; Paving, Tiling, Materials; Slating and Plastering; Practical Geometry, Mensuration, &c. By ADAM HAMMOND. Seventh Edition. 1s. 6d.
191. *PLUMBING.* A Text-Book to the Practice of the Art or Craft of the Plumber. With Chapters upon House Drainage and Ventilation. Fifth Edition. With 380 Illustrations. By W. P. BUCHAN. 3s. 6d.‡
192. *THE TIMBER IMPORTER'S, TIMBER MERCHANT'S,* and BUILDER'S STANDARD GUIDE. By R. E. GRANDY. 2s.
206. *A BOOK ON BUILDING, Civil and Ecclesiastical,* including CHURCH RESTORATION. With the Theory of Domes and the Great Pyramid, &c. By Sir EDMUND BECKETT, Bart., LL.D., Q.C., F.R.A.S. 4s. 6d.‡
226. *THE JOINTS MADE AND USED BY BUILDERS* in the Construction of various kinds of Engineering and Architectural Works. By WYVILL J. CHRISTY, Architect. With upwards of 160 Engravings on Wood. 3s.‡

☞ *The* ‡ *indicates that these vols. may be had strongly bound at* 6d. *extra.*

7, STATIONERS' HALL COURT, LUDGATE HILL, E.C.

WEALE'S RUDIMENTARY SERIES.

Architecture, Building, etc., *continued.*

228. *THE CONSTRUCTION OF ROOFS OF WOOD AND IRON.* By E. WYNDHAM TARN, M.A., Architect. Second Edition, revised. 1s. 6d.

229. *ELEMENTARY DECORATION:* as applied to the Interior and Exterior Decoration of Dwelling-Houses, &c. By J. W. FACEY. 2s.

257. *PRACTICAL HOUSE DECORATION.* A Guide to the Art of Ornamental Painting. By JAMES W. FACEY. 2s. 6d.

⁎⁎⁎ The two preceding Works, in One handsome Vol., half-bound, entitled "HOUSE DECORATION, ELEMENTARY AND PRACTICAL," *price* 5s.

230. *HANDRAILING.* Showing New and Simple Methods for finding the Pitch of the Plank. Drawing the Moulds, Bevelling, Jointing-up, and Squaring the Wreath. By GEORGE COLLINGS. Plates and Diagrams. 1s. 6d.

247. *BUILDING ESTATES:* a Rudimentary Treatise on the Development, Sale, Purchase, and General Management of Building Land. By FOWLER MAITLAND, Surveyor. Second Edition, revised. 2s.

248. *PORTLAND CEMENT FOR USERS.* By HENRY FAIJA, Assoc. M. Inst. C.E. Second Edition, corrected. Illustrated. 2s.

252. *BRICKWORK:* a Practical Treatise, embodying the General and Higher Principles of Bricklaying, Cutting and Setting, &c. By F. WALKER. Second Edition, Revised and Enlarged. 1s. 6d.

23. *THE PRACTICAL BRICK AND TILE BOOK.* Comprising:
189. BRICK AND TILE MAKING, by E. DOBSON, A.I.C.E.; PRACTICAL BRICKLAY-
252. ING, by A. HAMMOND; BRICKWORK, by F. WALKER. 550 pp. with 270 Illustrations. 6s. Strongly half-bound.

253. *THE TIMBER MERCHANT'S, SAW-MILLER'S, AND IMPORTER'S FREIGHT-BOOK AND ASSISTANT.* By WM. RICHARDSON. With a Chapter on Speeds of Saw-Mill Machinery, &c. By M. POWIS BALE, A.M.Inst.C.E. 3s.‡

258. *CIRCULAR WORK IN CARPENTRY AND JOINERY.* A Practical Treatise on Circular Work of Single and Double Curvature. By GEORGE COLLINGS, Author of "A Treatise on Handrailing." 2s. 6d.

259. *GAS FITTING:* A Practical Handbook treating of every Description of Gas Laying and Fitting. By JOHN BLACK. With 122 Illustrations. 2s. 6d.‡

261. *SHORING AND ITS APPLICATION:* A Handbook for the Use of Students. By GEORGE H. BLAGROVE. 1s. 6d. [*Just published.*

265. *THE ART OF PRACTICAL BRICK CUTTING & SETTING.* By ADAM HAMMOND. With 90 Engravings. 1s. 6d. [*Just published.*

267. *THE SCIENCE OF BUILDING:* An Elementary Treatise on the Principles of Construction. Adapted to the Requirements of Architectural Students. By E. WYNDHAM TARN, M.A. Lond. Third Edition, Revised and Enlarged. With 59 Wood Engravings. 3s. 6d.‡ [*Just published.*

SHIPBUILDING, NAVIGATION, MARINE ENGINEERING, ETC.

51. *NAVAL ARCHITECTURE.* An Exposition of the Elementary Principles of the Science, and their Practical Application to Naval Construction. By J. PEAKE. Fifth Edition, with Plates and Diagrams. 3s. 6d.‡

53*. *SHIPS FOR OCEAN & RIVER SERVICE,* Elementary and Practical Principles of the Construction of. By H. A. SOMMERFELDT. 1s. 6d.

53**. *AN ATLAS OF ENGRAVINGS* to Illustrate the above. Twelve large folding plates. Royal 4to, cloth. 7s. 6d.

54. *MASTING, MAST-MAKING, AND RIGGING OF SHIPS,* Also Tables of Spars, Rigging, Blocks; Chain, Wire, and Hemp Ropes, &c., relative to every class of vessels. By ROBERT KIPPING, N.A. 2s.

54*. *IRON SHIP-BUILDING.* With Practical Examples and Details. By JOHN GRANTHAM, C.E. 5th Edition. 4s.

☞ *The ‡ indicates that these vols. may be had strongly bound at 6d. extra.*

LONDON: CROSBY LOCKWOOD AND SON,

WEALE'S RUDIMENTARY SERIES. 7

Shipbuilding, Navigation, Marine Engineering, etc., *cont.*

55. *THE SAILOR'S SEA BOOK:* a Rudimentary Treatise on Navigation. By JAMES GREENWOOD, B.A. With numerous Woodcuts and Coloured Plates. New and enlarged edition. By W. H. ROSSER. 2s. 6d.‡
80. *MARINE ENGINES AND STEAM VESSELS.* By ROBERT MURRAY, C.E. Eighth Edition, thoroughly Revised, with Additions by the Author and by GEORGE CARLISLE, C.E., Senior Surveyor to the Board of Trade, Liverpool. 4s. 6d. limp; 5s. cloth boards.
83bis. *THE FORMS OF SHIPS AND BOATS.* By W. BLAND. Seventh Edition, Revised, with numerous Illustrations and Models. 1s. 6d.
99. *NAVIGATION AND NAUTICAL ASTRONOMY,* in Theory and Practice. By Prof. J. R. YOUNG. New Edition. 2s. 6d.
106. *SHIPS' ANCHORS,* a Treatise on. By G. COTSELL, N.A. 1s. 6d.
149. *SAILS AND SAIL-MAKING.* With Draughting, and the Centre of Effort of the Sails; Weights and Sizes of Ropes; Masting, Rigging, and Sails of Steam Vessels, &c. 12th Edition. By R. KIPPING. N.A., 2s. 6d.‡
155. *ENGINEER'S GUIDE TO THE ROYAL & MERCANTILE* NAVIES. By a PRACTICAL ENGINEER. Revised by D. F. M'CARTHY. 3s.
55 & 204. *PRACTICAL NAVIGATION.* Consisting of The Sailor's Sea-Book. By JAMES GREENWOOD and W. H. ROSSER. Together with the requisite Mathematical and Nautical Tables for the Working of the Problems. By H. LAW, C.E., and Prof. J. R. YOUNG. 7s. Half-bound.

AGRICULTURE, GARDENING, ETC.

61*. *A COMPLETE READY RECKONER FOR THE ADMEA-*SUREMENT OF LAND, &c. By A. ARMAN. Third Edition, revised and extended by C. NORRIS, Surveyor, Valuer, &c. 2s.
131. *MILLER'S, CORN MERCHANT'S, AND FARMER'S* READY RECKONER. Second Edition, with a Price List of Modern Flour-Mill Machinery, by W. S. HUTTON, C.E. 2s.
140. *SOILS, MANURES, AND CROPS.* (Vol. 1. OUTLINES OF MODERN FARMING.) By R. SCOTT BURN. Woodcuts. 2s.
141. *FARMING & FARMING ECONOMY,* Notes, Historical and Practical, on. (Vol. 2. OUTLINES OF MODERN FARMING.) By R. SCOTT BURN. 3s.
142. *STOCK; CATTLE, SHEEP, AND HORSES.* (Vol. 3. OUTLINES OF MODERN FARMING.) By R. SCOTT BURN. Woodcuts. 2s. 6d.
145. *DAIRY, PIGS, AND POULTRY,* Management of the. By R. SCOTT BURN. (Vol. 4. OUTLINES OF MODERN FARMING.) 2s.
146. *UTILIZATION OF SEWAGE, IRRIGATION, AND* RECLAMATION OF WASTE LAND. (Vol. 5. OUTLINES OF MODERN FARMING.) By R. SCOTT BURN. Woodcuts. 2s. 6d.
₊ Nos. 140-1-2-5-6, *in One Vol., handsomely half-bound, entitled* "OUTLINES OF MODERN FARMING." By ROBERT SCOTT BURN. *Price* 12s.
177. *FRUIT TREES,* The Scientific and Profitable Culture of. From the French of Du BREUIL. Revised by GEO. GLENNY. 187 Woodcuts. 3s. 6d.‡
198. *SHEEP; THE HISTORY, STRUCTURE, ECONOMY, AND* DISEASES OF. By W. C. SPOONER, M.R.V.C., &c. Fifth Edition, enlarged, including Specimens of New and Improved Breeds. 3s. 6d.‡
201. *KITCHEN GARDENING MADE EASY.* By GEORGE M. F. GLENNY. Illustrated. 1s. 6d.‡
207. *OUTLINES OF FARM MANAGEMENT, and the Organi-*zation of Farm Labour. By R. SCOTT BURN. 2s. 6d.‡
208. *OUTLINES OF LANDED ESTATES MANAGEMENT.* By R. SCOTT BURN. 2s. 6d.‡
₊ Nos. 207 & 208 *in One Vol., handsomely half-bound, entitled* "OUTLINES OF LANDED ESTATES AND FARM MANAGEMENT." By R. SCOTT BURN. *Price 6s.*

☞ *The ‡ indicates that these vols. may be had strongly bound at 6d. extra.*

7, STATIONERS' HALL COURT, LUDGATE HILL, E.C.

Agriculture, Gardening, etc., *continued.*

209. *THE TREE PLANTER AND PLANT PROPAGATOR.* A Practical Manual on the Propagation of Forest Trees, Fruit Trees, Flowering Shrubs, Flowering Plants, &c. By SAMUEL WOOD. 2s.‡

210. *THE TREE PRUNER.* A Practical Manual on the Pruning of Fruit Trees, including also their Training and Renovation; also the Pruning of Shrubs, Climbers, and Flowering Plants. By SAMUEL WOOD. 2s.‡

*** *Nos.* 209 *&* 210 *in One Vol., handsomely half-bound, entitled* "THE TREE PLANTER, PROPAGATOR, AND PRUNER." By SAMUEL WOOD. *Price* 5s.

218. *THE HAY AND STRAW MEASURER:* Being New Tables for the Use of Auctioneers, Valuers, Farmers, Hay and Straw Dealers, &c. By JOHN STEELE. Fourth Edition. 2s.

222. *SUBURBAN FARMING.* The Laying-out and Cultivation of Farms, adapted to the Produce of Milk, Butter, and Cheese, Eggs, Poultry, and Pigs. By Prof. JOHN DONALDSON and R. SCOTT BURN. 3s. 6d.‡

231. *THE ART OF GRAFTING AND BUDDING.* By CHARLES BALTET. With Illustrations. 2s. 6d.‡

232. *COTTAGE GARDENING;* or, Flowers, Fruits, and Vegetables for Small Gardens. By E. HOBDAY. 1s. 6d.

233. *GARDEN RECEIPTS.* Edited by CHARLES W. QUIN. 1s. 6d.

234. *MARKET AND KITCHEN GARDENING.* By C. W. SHAW, late Editor of "Gardening Illustrated." 3s.‡ [*Just published.*

239. *DRAINING AND EMBANKING.* A Practical Treatise, embodying the most recent experience in the Application of Improved Methods. By JOHN SCOTT, late Professor of Agriculture and Rural Economy at the Royal Agricultural College, Cirencester. With 68 Illustrations. 1s. 6d.

240. *IRRIGATION AND WATER SUPPLY.* A Treatise on Water Meadows, Sewage Irrigation, and Warping; the Construction of Wells, Ponds, and Reservoirs, &c. By Prof. JOHN SCOTT. With 34 Illus. 1s. 6d.

241. *FARM ROADS, FENCES, AND GATES.* A Practical Treatise on the Roads, Tramways, and Waterways of the Farm; the Principles of Enclosures; and the different kinds of Fences, Gates, and Stiles. By Professor JOHN SCOTT. With 75 Illustrations. 1s. 6d.

242. *FARM BUILDINGS.* A Practical Treatise on the Buildings necessary for various kinds of Farms, their Arrangement and Construction, with Plans and Estimates. By Prof. JOHN SCOTT. With 105 Illus. 2s.

243. *BARN IMPLEMENTS AND MACHINES.* A Practical Treatise on the Application of Power to the Operations of Agriculture; and on various Machines used in the Threshing-barn, in the Stock-yard, and in the Dairy, &c. By Prof. J. SCOTT. With 123 Illustrations. 2s.

244. *FIELD IMPLEMENTS AND MACHINES.* A Practical Treatise on the Varieties now in use, with Principles and Details of Construction, their Points of Excellence, and Management. By Professor JOHN SCOTT. With 138 Illustrations. 2s.

245. *AGRICULTURAL SURVEYING.* A Practical Treatise on Land Surveying, Levelling, and Setting-out; and on Measuring and Estimating Quantities, Weights, and Values of Materials, Produce, Stock, &c. By Prof. JOHN SCOTT. With 62 Illustrations. 1s. 6d.

*** *Nos.* 239 *to* 245 *in One Vol., handsomely half-bound, entitled* "THE COMPLETE TEXT-BOOK OF FARM ENGINEERING." By Professor JOHN SCOTT. *Price* 12s.

250. *MEAT PRODUCTION.* A Manual for Producers, Distributors, &c. By JOHN EWART. 2s. 6d.‡

266. *BOOK-KEEPING FOR FARMERS & ESTATE OWNERS.* By J. M. WOODMAN, Chartered Accountant. 2s. 6d. cloth limp; 3s. 6d. cloth boards. [*Just published.*

☞ *The* ‡ *indicates that these vols. may be had strongly bound at* 6d. *extra.*

LONDON : CROSBY LOCKWOOD AND SON,

MATHEMATICS, ARITHMETIC, ETC.

32. *MATHEMATICAL INSTRUMENTS*, a Treatise on; Their Construction, Adjustment, Testing, and Use concisely Explained. By J. F. HEATHER, M.A. Fourteenth Edition, revised, with additions, by A. T. WALMISLEY, M.I.C.E., Fellow of the Surveyors' Institution. Original Edition, in 1 vol., Illustrated. 2s. ‡ [*Just published.*

⁎ In ordering the above, be careful to say, "*Original Edition*" (*No.* 32), to distinguish it from the Enlarged Edition in 3 vols. (*Nos.* 168-9-70.)

76. *DESCRIPTIVE GEOMETRY*, an Elementary Treatise on; with a Theory of Shadows and of Perspective, extracted from the French of G. MONGE. To which is added, a description of the Principles and Practice of Isometrical Projection. By J. F. HEATHER, M.A. With 14 Plates. 2s.

178. *PRACTICAL PLANE GEOMETRY:* giving the Simplest Modes of Constructing Figures contained in one Plane and Geometrical Construction of the Ground. By J. F. HEATHER, M.A. With 215 Woodcuts. 2s.

83. *COMMERCIAL BOOK-KEEPING.* With Commercial Phrases and Forms in English, French, Italian, and German. By JAMES HADDON, M.A., Arithmetical Master of King's College School, London. 1s. 6d.

84. *ARITHMETIC*, a Rudimentary Treatise on: with full Explanations of its Theoretical Principles, and numerous Examples for Practice. By Professor J. R. YOUNG. Eleventh Edition. 1s. 6d.

84*. A KEY to the above, containing Solutions in full to the Exercises, together with Comments, Explanations, and Improved Processes, for the Use of Teachers and Unassisted Learners. By J. R. YOUNG. 1s. 6d.

85. *EQUATIONAL ARITHMETIC*, applied to Questions of Interest, Annuities, Life Assurance, and General Commerce; with various Tables by which all Calculations may be greatly facilitated. By W. HIPSLEY. 2s.

86. *ALGEBRA*, the Elements of. By JAMES HADDON, M.A. With Appendix, containing miscellaneous Investigations, and a Collection of Problems in various parts of Algebra. 2s.

86*. A KEY AND COMPANION to the above Book, forming an extensive repository of Solved Examples and Problems in Illustration of the various Expedients necessary in Algebraical Operations. By J. R. YOUNG. 1s. 6d.

88. *EUCLID*, THE ELEMENTS OF: with many additional Propositions
89. and Explanatory Notes: to which is prefixed, an Introductory Essay on Logic. By HENRY LAW, C.E. 2s. 6d. ‡

⁎ *Sold also separately, viz.:—*

88. EUCLID, The First Three Books. By HENRY LAW, C.E. 1s. 6d.
89. EUCLID, Books 4, 5, 6, 11, 12. By HENRY LAW, C.E. 1s. 6d.

90. *ANALYTICAL GEOMETRY AND CONIC SECTIONS*, By JAMES HANN. A New Edition, by Professor J. R. YOUNG. 2s. ‡

91. *PLANE TRIGONOMETRY*, the Elements of. By JAMES HANN, formerly Mathematical Master of King's College, London. 1s. 6d.

92. *SPHERICAL TRIGONOMETRY*, the Elements of. By JAMES HANN. Revised by CHARLES H. DOWLING, C.E. 1s.
⁎ *Or with "The Elements of Plane Trigonometry," in One Volume,* 2s. 6d.

93. *MENSURATION AND MEASURING.* With the Mensuration and Levelling of Land for the Purposes of Modern Engineering. By T. BAKER, C.E. New Edition by E. NUGENT, C.E. Illustrated. 1s. 6d.

101. *DIFFERENTIAL CALCULUS*, Elements of the. By W. S. B. WOOLHOUSE, F.R.A.S., &c. 1s. 6d.

102. *INTEGRAL CALCULUS*, Rudimentary Treatise on the. By HOMERSHAM COX, B.A. Illustrated. 1s.

136. *ARITHMETIC*, Rudimentary, for the Use of Schools and Self-Instruction. By JAMES HADDON, M.A. Revised by A. ARMAN. 1s. 6d.

137. A KEY TO HADDON'S RUDIMENTARY ARITHMETIC. By A. ARMAN. 1s. 6d.

☞ *The ‡ indicates that these vols. may be had strongly bound at 6d. extra.*

7, STATIONERS' HALL COURT, LUDGATE HILL, E.C.

Mathematics, Arithmetic, etc., *continued.*

168. *DRAWING AND MEASURING INSTRUMENTS.* Including—I. Instruments employed in Geometrical and Mechanical Drawing, and in the Construction, Copying, and Measurement of Maps and Plans. II. Instruments used for the purposes of Accurate Measurement, and for Arithmetical Computations. By J. F. HEATHER, M.A. Illustrated. 1s. 6d

169. *OPTICAL INSTRUMENTS.* Including (more especially) Telescopes, Microscopes, and Apparatus for producing copies of Maps and Plans by Photography. By J. F. HEATHER, M.A. Illustrated. 1s. 6d.

170. *SURVEYING AND ASTRONOMICAL INSTRUMENTS.* Including—I. Instruments Used for Determining the Geometrical Features of a portion of Ground. II. Instruments Employed in Astronomical Observations. By J. F. HEATHER, M.A. Illustrated. 1s. 6d.

⁎⁎ *The above three volumes form an enlargement of the Author's original work "Mathematical Instruments." (See No. 32 in the Series.)*

168.⎫
169.⎬ *MATHEMATICAL INSTRUMENTS.* By J. F. HEATHER, M.A. Enlarged Edition, for the most part entirely re-written. The 3 Parts as
170.⎭ above, in One thick Volume. With numerous Illustrations. 4s. 6d.‡

158. *THE SLIDE RULE, AND HOW TO USE IT;* containing full, easy, and simple Instructions to perform all Business Calculations with unexampled rapidity and accuracy. By CHARLES HOARE, C.E. Fifth Edition. With a Slide Rule in tuck of cover. 2s. 6d.‡

196. *THEORY OF COMPOUND INTEREST AND ANNUITIES;* with Tables of Logarithms for the more Difficult Computations of Interest, Discount, Annuities, &c. By FÉDOR THOMAN. 4s.‡

199. *THE COMPENDIOUS CALCULATOR;* or, Easy and Concise Methods of Performing the various Arithmetical Operations required in Commercial and Business Transactions; together with Useful Tables. By D. O'GORMAN. Twenty-seventh Edition, carefully revised by C. NORRIS. 2s. 6d., cloth limp; 3s. 6d., strongly half-bound in leather.

204. *MATHEMATICAL TABLES,* for Trigonometrical, Astronomical, and Nautical Calculations; to which is prefixed a Treatise on Logarithms. By HENRY LAW, C.E. Together with a Series of Tables for Navigation and Nautical Astronomy. By Prof. J. R. YOUNG. New Edition. 4s.

204*. *LOGARITHMS.* With Mathematical Tables for Trigonometrical, Astronomical, and Nautical Calculations. By HENRY LAW, M.Inst.C.E. New and Revised Edition. (Forming part of the above Work). 3s.

221. *MEASURES, WEIGHTS, AND MONEYS OF ALL NATIONS,* and an Analysis of the Christian, Hebrew, and Mahometan Calendars. By W. S. B. WOOLHOUSE, F.R.A.S., F.S.S. Sixth Edition. 2s.‡

227. *MATHEMATICS AS APPLIED TO THE CONSTRUCTIVE ARTS.* Illustrating the various processes of Mathematical Investigation, by means of Arithmetical and Simple Algebraical Equations and Practical Examples. By FRANCIS CAMPIN. C.E. Second Edition. 3s.‡

PHYSICAL SCIENCE, NATURAL PHILOSOPHY, ETC.

1. *CHEMISTRY.* By Professor GEORGE FOWNES, F.R.S. With an Appendix on the Application of Chemistry to Agriculture. 1s.

2. *NATURAL PHILOSOPHY,* Introduction to the Study of. By C. TOMLINSON. Woodcuts. 1s. 6d.

6. *MECHANICS,* Rudimentary Treatise on. By CHARLES TOMLINSON. Illustrated. 1s. 6d.

7. *ELECTRICITY;* showing the General Principles of Electrical Science, and the purposes to which it has been applied. By Sir W. SNOW HARRIS, F.R.S., &c. With Additions by R. SABINE, C.E., F.S.A. 1s. 6d.

7*. *GALVANISM.* By Sir W. SNOW HARRIS. New Edition by ROBERT SABINE, C.E., F.S.A. 1s. 6d.

8. *MAGNETISM;* being a concise Exposition of the General Principles of Magnetical Science. By Sir W. SNOW HARRIS. New Edition, revised by H. M. NOAD, Ph.D. With 165 Woodcuts. 3s. 6d.‡

☞ *The ‡ indicates that these vols. may be had strongly bound at 6d. extra*

Physical Science, Natural Philosophy, etc., *continued.*

11. *THE ELECTRIC TELEGRAPH;* its History and Progress; with Descriptions of some of the Apparatus. By R. SABINE, C.E., F.S.A. 3s.
12. *PNEUMATICS,* including Acoustics and the Phenomena of Wind Currents, for the Use of Beginners. By CHARLES TOMLINSON, F.R.S. Fourth Edition, enlarged. Illustrated. 1s. 6d. [*Just published.*
72. *MANUAL OF THE MOLLUSCA;* a Treatise on Recent and Fossil Shells. By Dr. S. P. WOODWARD, A.L.S. Fourth Edition. With Appendix by RALPH TATE, A.L.S., F.G.S. With numerous Plates and 300 Woodcuts. 6s. 6d. Cloth boards, 7s. 6d.
96. *ASTRONOMY.* By the late Rev. ROBERT MAIN, M.A. Third Edition, by WILLIAM THYNNE LYNN, B.A., F.R.A.S. 2s.
97. *STATICS AND DYNAMICS,* the Principles and Practice of; embracing also a clear development of Hydrostatics, Hydrodynamics, and Central Forces. By T. BAKER, C.E. Fourth Edition. 1s. 6d.
138. *TELEGRAPH,* Handbook of the; a Guide to Candidates for Employment in the Telegraph Service. By R. BOND. 3s.‡
173. *PHYSICAL GEOLOGY,* partly based on Major-General PORTLOCK's "Rudiments of Geology." By RALPH TATE, A.L.S., &c. Woodcuts. 2s.
174. *HISTORICAL GEOLOGY,* partly based on Major-General PORTLOCK's "Rudiments." By RALPH TATE, A.L.S., &c. Woodcuts. 2s. 6d.
173 *RUDIMENTARY TREATISE ON GEOLOGY,* Physical and
& Historical. Partly based on Major-General PORTLOCK's "Rudiments of
174. Geology." By RALPH TATE, A.L.S., F.G.S., &c. In One Volume. 4s. 6d.‡
183 *ANIMAL PHYSICS,* Handbook of. By Dr. LARDNER, D.C.L.,
& formerly Professor of Natural Philosophy and Astronomy in University
184. College, Lond. With 520 Illustrations. In One Vol. 7s. 6d., cloth boards.
*** *Sold also in Two Parts, as follows :—*
183. ANIMAL PHYSICS. By Dr. LARDNER. Part I., Chapters I.—VII. 4s.
184. ANIMAL PHYSICS. By Dr. LARDNER. Part II., Chapters VIII.—XVIII. 3s.

FINE ARTS.

20. *PERSPECTIVE FOR BEGINNERS.* Adapted to Young Students and Amateurs in Architecture, Painting, &c. By GEORGE PYNE. 2s.
40 *GLASS STAINING, AND THE ART OF PAINTING ON GLASS.* From the German of Dr. GESSERT and EMANUEL OTTO FROMBERG. With an Appendix on THE ART OF ENAMELLING. 2s. 6d.
69. *MUSIC,* A Rudimentary and Practical Treatise on. With numerous Examples. By CHARLES CHILD SPENCER. 2s. 6d.
71. *PIANOFORTE,* The Art of Playing the. With numerous Exercises & Lessons from the Best Masters. By CHARLES CHILD SPENCER. 1s. 6d.
69-71. *MUSIC & THE PIANOFORTE.* In one vol. Half bound, 5s.
181. *PAINTING POPULARLY EXPLAINED,* including Fresco, Oil, Mosaic, Water Colour, Water-Glass, Tempera, Encaustic, Miniature, Painting on Ivory, Vellum, Pottery, Enamel, Glass, &c. With Historical Sketches of the Progress of the Art by THOMAS JOHN GULLICK, assisted by JOHN TIMBS, F.S.A. Fifth Edition, revised and enlarged. 5s.‡
186. *A GRAMMAR OF COLOURING,* applied to Decorative Painting and the Arts. By GEORGE FIELD. New Edition, enlarged and adapted to the Use of the Ornamental Painter and Designer. By ELLIS A. DAVIDSON. With two new Coloured Diagrams, &c. 3s.‡
246. *A DICTIONARY OF PAINTERS, AND HANDBOOK FOR PICTURE AMATEURS;* including Methods of Painting, Cleaning, Relining and Restoring, Schools of Painting, &c. With Notes on the Copyists and Imitators of each Master. By PHILIPPE DARYL. 2s. 6d.‡

☞ *The ‡ indicates that these vols. may be had strongly bound at 6d. extra.*

INDUSTRIAL AND USEFUL ARTS.

23. *BRICKS AND TILES*, Rudimentary Treatise on the Manufacture of. By E. Dobson, M.R.I.B.A. Illustrated, 3s.‡
67. *CLOCKS, WATCHES, AND BELLS*, a Rudimentary Treatise on. By Sir Edmund Beckett, LL.D., Q.C. Seventh Edition, revised and enlarged. 4s. 6d. limp; 5s. 6d. cloth boards.
83**. *CONSTRUCTION OF DOOR LOCKS*. Compiled from the Papers of A. C. Hobbs, and Edited by Charles Tomlinson. F.R.S. 2s. 6d.
162. *THE BRASS FOUNDER'S MANUAL;* Instructions for Modelling, Pattern-Making, Moulding, Turning, Filing, Burnishing, Bronzing, &c. With copious Receipts, &c. By Walter Graham. 2s.‡
205. *THE ART OF LETTER PAINTING MADE EASY*. By J.G. Badenoch. Illustrated with 12 full-page Engravings of Examples. 1s. 6d.
215. *THE GOLDSMITH'S HANDBOOK*, containing full Instructions for the Alloying and Working of Gold. By George E. Gee, 3s.‡
225. *THE SILVERSMITH'S HANDBOOK*, containing full Instructions for the Alloying and Working of Silver. By George E. Gee. 3s.‡
₊ The two preceding Works, in One handsome Vol., half-bound, entitled "The Goldsmith's & Silversmith's Complete Handbook," 7s.
249. *THE HALL-MARKING OF JEWELLERY PRACTICALLY CONSIDERED*. By George E. Gee. 3s.‡
224. *COACH BUILDING*, A Practical Treatise, Historical and Descriptive. By J. W. Burgess. 2s. 6d.‡
235. *PRACTICAL ORGAN BUILDING*. By W. E. Dickson, M.A., Precentor of Ely Cathedral. Illustrated. 2s. 6d.‡
262. *THE ART OF BOOT AND SHOEMAKING*, including Measurement, 'Last-fitting, Cutting-out, Closing and Making. By John Bedford Leno. Numerous Illustrations. Third Edition. 2s.
263. *MECHANICAL DENTISTRY:* A Practical Treatise on the Construction of the Various Kinds of Artificial Dentures, with Formulæ, Tables, Receipts, &c. By Charles Hunter. Third Edition. 3s.‡

MISCELLANEOUS VOLUMES.

36. *A DICTIONARY OF TERMS used in ARCHITECTURE, BUILDING, ENGINEERING, MINING, METALLURGY, ARCHÆOLOGY, the FINE ARTS, &c*. By John Weale. Fifth Edition. Revised by Robert Hunt, F.R.S. Illustrated. 5s. limp; 6s. cloth boards.
50. *THE LAW OF CONTRACTS FOR WORKS AND SERVICES*. By David Gibbons. Third Edition, enlarged. 3s.‡
112. *MANUAL OF DOMESTIC MEDICINE*. By R. Gooding, B.A., M.D. A Family Guide in all Cases of Accident and Emergency. 2s.‡
112*. *MANAGEMENT OF HEALTH*. A Manual of Home and Personal Hygiene. By the Rev. James Baird, B.A. 1s.
150. *LOGIC*, Pure and Applied. By S. H. Emmens. 1s. 6d.
153. *SELECTIONS FROM LOCKE'S ESSAYS ON THE HUMAN UNDERSTANDING*. With Notes by S. H. Emmens. 2s.
154. *GENERAL HINTS TO EMIGRANTS*. 2s.
157. *THE EMIGRANT'S GUIDE TO NATAL*. By Robert James Mann, F.R.A.S., F.M.S. Second Edition. Map. 2s.
193. *HANDBOOK OF FIELD FORTIFICATION*. By Major W. W. Knollys, F.R.G.S. With 163 Woodcuts. 3s.‡
194. *THE HOUSE MANAGER:* Being a Guide to Housekeeping. Practical Cookery, Pickling and Preserving, Household Work, Dairy Management, &c. By An Old Housekeeper. 3s. 6d.‡
194, 112 & 112*. *HOUSE BOOK (The)*. Comprising:—I. The House Manager. By an Old Housekeeper. II. Domestic Medicine. By R. Gooding, M.D. III. Management of Health. By J. Baird. In One Vol., half-bound, 6s.

☞ The ‡ indicates that these vols. may be had strongly bound at 6d. extra.

LONDON : CROSBY LOCKWOOD AND SON.

EDUCATIONAL AND CLASSICAL SERIES.

HISTORY.

1. **England, Outlines of the History of;** more especially with reference to the Origin and Progress of the English Constitution. By WILLIAM DOUGLAS HAMILTON, F.S.A., of Her Majesty's Public Record Office. 4th Edition, revised. 5s.; cloth boards, 6s.
5. **Greece, Outlines of the History of;** in connection with the Rise of the Arts and Civilization in Europe. By W. DOUGLAS HAMILTON, of University College, London, and EDWARD LEVIEN, M.A., of Balliol College, Oxford. 2s. 6d.; cloth boards, 3s. 6d.
7. **Rome, Outlines of the History of:** from the Earliest Period to the Christian Era and the Commencement of the Decline of the Empire. By EDWARD LEVIEN, of Balliol College, Oxford. Map, 2s. 6d.; cl. bds. 3s. 6d.
9. **Chronology of History, Art, Literature, and Progress,** from the Creation of the World to the Present Time. The Continuation by W. D. HAMILTON, F.S.A. 3s.; cloth boards, 3s. 6d.
50. **Dates and Events in English History,** for the use of Candidates in Public and Private Examinations. By the Rev. E. RAND. 1s.

ENGLISH LANGUAGE AND MISCELLANEOUS.

11. **Grammar of the English Tongue,** Spoken and Written. With an Introduction to the Study of Comparative Philology. By HYDE CLARKE, D.C.L. Fourth Edition. 1s. 6d.
12. **Dictionary of the English Language,** as Spoken and Written. Containing above 100,000 Words. By HYDE CLARKE, D.C.L. 3s. 6d.; cloth boards, 4s. 6d.; complete with the GRAMMAR, cloth bds., 5s. 6d.
48. **Composition and Punctuation,** familiarly Explained for those who have neglected the Study of Grammar. By JUSTIN BRENAN. 18th Edition. 1s. 6d.
49. **Derivative Spelling-Book:** Giving the Origin of Every Word from the Greek, Latin, Saxon, German, Teutonic, Dutch, French, Spanish, and other Languages; with their present Acceptation and Pronunciation. By J. ROWBOTHAM, F.R.A.S. Improved Edition. 1s. 6d.
51. **The Art of Extempore Speaking:** Hints for the Pulpit, the Senate, and the Bar. By M. BAUTAIN, Vicar-General and Professor at the Sorbonne. Translated from the French. 8th Edition, carefully corrected. 2s. 6d.
53. **Places and Facts in Political and Physical Geography,** for Candidates in Examinations. By the Rev. EDGAR RAND, B.A. 1s.
54. **Analytical Chemistry,** Qualitative and Quantitative, a Course of. To which is prefixed, a Brief Treatise upon Modern Chemical Nomenclature and Notation. By WM. W. PINK and GEORGE E. WEBSTER. 2s.

THE SCHOOL MANAGERS' SERIES OF READING BOOKS,

Edited by the Rev. A. R. GRANT, Rector of Hitcham, and Honorary Canon of Ely; formerly H.M. Inspector of Schools.
INTRODUCTORY PRIMER, 3d.

	s. d.		s. d
FIRST STANDARD	0 6	FOURTH STANDARD	1 2
SECOND ,,	0 10	FIFTH ,,	1 6
THIRD ,,	1 0	SIXTH ,,	1 6

LESSONS FROM THE BIBLE. Part I. Old Testament. 1s.
LESSONS FROM THE BIBLE. Part II. New Testament, to which is added THE GEOGRAPHY OF THE BIBLE, for very young Children. By Rev. C. THORNTON FORSTER. 1s. 2d. *_** Or the Two Parts in One Volume. 2s.

7, STATIONERS' HALL COURT, LUDGATE HILL, E.C.

FRENCH.

24. **French Grammar.** With Complete and Concise Rules on the Genders of French Nouns. By G. L. STRAUSS, Ph.D. 1s. 6d.
25. **French-English Dictionary.** Comprising a large number of New Terms used in Engineering, Mining, &c. By ALFRED ELWES. 1s. 6d.
26. **English-French Dictionary.** By ALFRED ELWES. 2s.
25,26. **French Dictionary** (as above). Complete, in One Vol., 3s.; cloth boards, 3s. 6d. *** Or with the GRAMMAR, cloth boards, 4s. 6d.
47. **French and English Phrase Book:** containing Introductory Lessons, with Translations, several Vocabularies of Words, a Collection of suitable Phrases, and Easy Familiar Dialogues. 1s. 6d.

GERMAN.

39. **German Grammar.** Adapted for English Students, from Heyse's Theoretical and Practical Grammar, by Dr. G. L. STRAUSS. 1s. 6d.
40. **German Reader:** A Series of Extracts, carefully culled from the most approved Authors of Germany; with Notes, Philological and Explanatory. By G. L. STRAUSS, Ph.D. 1s.
41-43. **German Triglot Dictionary.** By N. E. S. A. HAMILTON. In Three Parts. Part I. German-French-English. Part II. English-German-French. Part III. French-German-English. 3s., or cloth boards, 4s.
41-43 & 39. **German Triglot Dictionary** (as above), together with German Grammar (No. 39), in One Volume, cloth boards, 5s.

ITALIAN.

27. **Italian Grammar,** arranged in Twenty Lessons, with a Course of Exercises. By ALFRED ELWES. 1s. 6d.
28. **Italian Triglot Dictionary,** wherein the Genders of all the Italian and French Nouns are carefully noted down. By ALFRED ELWES. Vol. 1. Italian-English-French. 2s. 6d.
30. **Italian Triglot Dictionary.** By A. ELWES. Vol. 2. English-French-Italian. 2s. 6d.
32. **Italian Triglot Dictionary.** By ALFRED ELWES. Vol. 3. French-Italian-English. 2s. 6d.
28,30, **Italian Triglot Dictionary** (as above). In One Vol., 7s. 6d
32. Cloth boards.

SPANISH AND PORTUGUESE.

34. **Spanish Grammar,** in a Simple and Practical Form. With a Course of Exercises. By ALFRED ELWES. 1s. 6d.
35. **Spanish-English and English-Spanish Dictionary.** Including a large number of Technical Terms used in Mining, Engineering, &c. with the proper Accents and the Gender of every Noun. By ALFRED ELWES 4s.; cloth boards, 5s. *** Or with the GRAMMAR, cloth boards, 6s.
55. **Portuguese Grammar,** in a Simple and Practical Form. With a Course of Exercises. By ALFRED ELWES. 1s. 6d.
56. **Portuguese-English and English-Portuguese Dictionary.** Including a large number of Technical Terms used in Mining, Engineering, &c., with the proper Accents and the Gender of every Noun. By ALFRED ELWES. Second Edition, Revised, 5s.; cloth boards, 6s. *** Or with the GRAMMAR, cloth boards, 7s.

HEBREW.

46*. **Hebrew Grammar.** By Dr. BRESSLAU. 1s. 6d.
44. **Hebrew and English Dictionary,** Biblical and Rabbinical; containing the Hebrew and Chaldee Roots of the Old Testament Post-Rabbinical Writings. By Dr. BRESSLAU. 6s.
46. **English and Hebrew Dictionary.** By Dr. BRESSLAU. 3s.
44,46. **Hebrew Dictionary** (as above), in Two Vols., complete, with
46*. the GRAMMAR, cloth boards, 12s.

LONDON: CROSBY LOCKWOOD AND SON.

LATIN.

19. **Latin Grammar.** Containing the Inflections and Elementary Principles of Translation and Construction. By the Rev. THOMAS GOODWIN, M.A., Head Master of the Greenwich Proprietary School. 1s. 6d.
20. **Latin-English Dictionary.** By the Rev. THOMAS GOODWIN, M.A. 2s.
22. **English-Latin Dictionary;** together with an Appendix of French and Italian Words which have their origin from the Latin. By the Rev. THOMAS GOODWIN, M.A. 1s. 6d.
20,22. **Latin Dictionary** (as above). Complete in One Vol., 3s. 6d. cloth boards, 4s. 6d. *** Or with the GRAMMAR, cloth boards, 5s. 6d.

LATIN CLASSICS. With Explanatory Notes in English.

1. **Latin Delectus.** Containing Extracts from Classical Authors, with Genealogical Vocabularies and Explanatory Notes, by H. YOUNG. 1s. 6d.
2. **Cæsaris Commentarii de Bello Gallico.** Notes, and a Geographical Register for the Use of Schools, by H. YOUNG. 2s.
3. **Cornelius Nepos.** With Notes. By H. YOUNG. 1s.
4. **Virgilii Maronis Bucolica et Georgica.** With Notes on the Bucolics by W. RUSHTON, M.A., and on the Georgics by H. YOUNG. 1s. 6d.
5. **Virgilii Maronis Æneis.** With Notes, Critical and Explanatory, by H. YOUNG. New Edition, revised and improved With copious Additional Notes by Rev. T. H. L. LEARY, D.C.L., formerly Scholar of Brasenose College, Oxford. 3s.
5*. ——— Part 1. Books i.—vi., 1s. 6d.
5**. ——— Part 2. Books vii.—xii., 2s.
6. **Horace;** Odes, Epode, and Carmen Sæculare. Notes by H. YOUNG. 1s. 6d.
7. **Horace;** Satires, Epistles, and Ars Poetica. Notes by W. BROWNRIGG SMITH, M.A., F.R.G.S. 1s. 6d.
8. **Sallustii Crispi Catalina et Bellum Jugurthinum.** Notes, Critical and Explanatory, by W. M. DONNE, B.A., Trin. Coll., Cam. 1s. 6d.
9. **Terentii Andria et Heautontimorumenos.** With Notes, Critical and Explanatory, by the Rev. JAMES DAVIES, M.A. 1s. 6d.
10. **Terentii Adelphi, Hecyra, Phormio.** Edited, with Notes, Critical and Explanatory, by the Rev. JAMES DAVIES, M.A. 2s.
11. **Terentii Eunuchus, Comœdia.** Notes, by Rev. J. DAVIES, M.A. 1s. 6d.
12. **Ciceronis Oratio pro Sexto Roscio Amerino.** Edited, with an Introduction, Analysis, and Notes, Explanatory and Critical, by the Rev JAMES DAVIES, M.A. 1s. 6d.
13. **Ciceronis Orationes in Catilinam, Verrem, et pro Archia.** With Introduction, Analysis, and Notes, Explanatory and Critical, by Rev. T. H. L. LEARY, D.C.L. formerly Scholar of Brasenose College, Oxford. 1s. 6d.
14. **Ciceronis Cato Major, Lælius, Brutus, sive de Senectute, de Amicitia, de Claris Oratoribus Dialogi.** With Notes by W. BROWNRIGG SMITH M.A., F.R.G.S. 2s.
16. **Livy;** History of Rome. Notes by H. YOUNG and W. B. SMITH, M.A. Part 1. Books i., ii., 1s. 6d.
16*. ——— Part 2. Books iii., iv., v., 1s. 6d.
17. ——— Part 3. Books xxi., xxii., 1s. 6d.
19. **Latin Verse Selections,** from Catullus, Tibullus, Propertius, and Ovid. Notes by W. B. DONNE, M.A., Trinity College, Cambridge. 2s.
20. **Latin Prose Selections,** from Varro, Columella, Vitruvius, Seneca, Quintilian, Florus, Velleius Paterculus, Valerius Maximus Suetonius, Apuleius, &c. Notes by W. B. DONNE, M.A. 2s.
21. **Juvenalis Satiræ.** With Prolegomena and Notes by T. H. S. ESCOTT, B.A., Lecturer on Logic at King's College, London. 2s.

7. STATIONERS' HALL COURT. LUDGATE HILL, E.C.

GREEK.

14. **Greek Grammar,** in accordance with the Principles and Philological Researches of the most eminent Scholars of our own day. By HANS CLAUDE HAMILTON. 1s. 6d.
15, 17. **Greek Lexicon.** Containing all the Words in General Use, with their Significations, Inflections, and Doubtful Quantities. By HENRY R. HAMILTON. Vol. 1. Greek-English, 2s. 6d.; Vol. 2. English-Greek, 2s. Or the Two Vols. in One, 4s. 6d.: cloth boards, 5s.
14, 15, 17. **Greek Lexicon** (as above). Complete, with the GRAMMAR, in One Vol., cloth boards, 6s.

GREEK CLASSICS. With Explanatory Notes in English.

1. **Greek Delectus.** Containing Extracts from Classical Authors, with Genealogical Vocabularies and Explanatory Notes, by H. YOUNG. New Edition, with an improved and enlarged Supplementary Vocabulary, by JOHN HUTCHISON, M.A., of the High School, Glasgow. 1s. 6d.
2, 3. **Xenophon's Anabasis;** or, The Retreat of the Ten Thousand. Notes and a Geographical Register, by H. YOUNG. Part 1. Books i. to iii., 1s. Part 2. Books iv. to vii., 1s.
4. **Lucian's Select Dialogues.** The Text carefully revised, with Grammatical and Explanatory Notes, by H. YOUNG. 1s. 6d.
5-12. **Homer, The Works of.** According to the Text of BAEUMLEIN. With Notes, Critical and Explanatory, drawn from the best and latest Authorities, with Preliminary Observations and Appendices, by T. H. L. LEARY, M.A., D.C.L.
  THE ILIAD: Part 1. Books i. to vi., 1s. 6d. | Part 3. Books xiii. to xviii., 1s. 6d.
  Part 2. Books vii. to xii., 1s. 6d. | Part 4. Books xix. to xxiv., 1s. 6d.
  THE ODYSSEY: Part 1. Books i. to vi., 1s. 6d | Part 3. Books xiii. to xviii., 1s. 6d.
  Part 2. Books vii. to xii., 1s. 6d. | Part 4. Books xix. to xxiv., and Hymns, 2s.
13. **Plato's Dialogues:** The Apology of Socrates, the Crito, and the Phædo. From the Text of C. F. HERMANN. Edited with Notes, Critical and Explanatory, by the Rev. JAMES DAVIES, M.A. 2s.
14-17. **Herodotus, The History of,** chiefly after the Text of GAISFORD. With Preliminary Observations and Appendices, and Notes, Critical and Explanatory, by T. H. L. LEARY, M.A., D.C.L.
  Part 1. Books i., ii. (The Clio and Euterpe), 2s.
  Part 2. Books iii., iv. (The Thalia and Melpomene), 2s.
  Part 3. Books v.-vii. (The Terpsichore, Erato, and Polymnia), 2s.
  Part 4. Books viii., ix. (The Urania and Calliope) and Index, 1s. 6d.
18. **Sophocles:** Œdipus Tyrannus. Notes by H. YOUNG. 1s.
20. **Sophocles:** Antigone. From the Text of DINDORF. Notes, Critical and Explanatory, by the Rev. JOHN MILNER, B.A. 2s.
23. **Euripides:** Hecuba and Medea. Chiefly from the Text of DINDORF. With Notes, Critical and Explanatory, by W. BROWNRIGG SMITH, M.A., F.R.G.S. 1s. 6d.
26. **Euripides:** Alcestis. Chiefly from the Text of DINDORF. With Notes, Critical and Explanatory, by JOHN MILNER, B.A. 1s. 6d.
30. **Æschylus:** Prometheus Vinctus: The Prometheus Bound. From the Text of DINDORF. Edited, with English Notes, Critical and Explanatory, by the Rev. JAMES DAVIES, M.A. 1s.
32. **Æschylus:** Septem Contra Thebes: The Seven against Thebes. From the Text of DINDORF. Edited, with English Notes, Critical and Explanatory, by the Rev. JAMES DAVIES, M.A. 1s.
40. **Aristophanes:** Acharnians. Chiefly from the Text of C. H. WEISE. With Notes, by C. S. T. TOWNSHEND, M.A. 1s. 6d.
41. **Thucydides:** History of the Peloponnesian War. Notes by H. YOUNG. Book 1. 1s. 6d.
42. **Xenophon's Panegyric on Agesilaus.** Notes and Introduction by LL. F. W. JEWITT. 1s. 6d.
43. **Demosthenes.** The Oration on the Crown and the Philippics. With English Notes. By Rev. T. H. L. LEARY, D.C.L., formerly Scholar of Brasenose College, Oxford. 1s. 6d.

www.ingramcontent.com/pod-product-compliance
Lightning Source LLC
Chambersburg PA
CBHW030309170426
43202CB00009B/930